MSRP - $24.95

AC6V's

Amateur Radio and DX Reference Guide

DX101X
HF + 6 METERS
DXING REFERENCE GUIDE

By Rodney R. Dinkins AC6V

AC6V Publications
981 Texas Rd
Iola, KS
66749

PREFACE

Picture a nerdy high school kid feverishly stringing antennas up and down a steep - roofed two-story house in Canton, Ohio. Then further picture this kid staying up nights listening to WGN Chicago and WOR New York through WWII surplus headphones. The years were 1947-1950 and I had been given an old FADA AM radio by an elmer. Little did I know that I had been bitten by the DX bug. Del Rio TX, Council Bluffs IA, Nashville TN all entered my logbook as I tuned the BC band. Subsequently, I built crystal sets, and a 3–tube Knight-kit Ocean Hopper. Later I saved for many months (at 50 cents an hour) to buy a Hallicrafters S-40A for $89.95. Then I could hear the Amateur Bands and knew that was for me.

After a stint in US Naval Aviation (1951-1955) where I could legitimately say Roger, Wilco, and Out and operate Morse Code on Pacific overseas flights, it was not until 1977 that I finally decided to get a Ham ticket, and the DX bug was still ingrained in my noggin. I still work DX (made Honor Roll 3 years ago) and enjoy maintaining the AC6V Website (over 10 million hits). As a result of the website and our local repeater, I found that many newer Amateurs had numerous questions about DXing and encouraged me to write a book on the subject. In September, 2004, six meter DXing has been added.

This book has been a year in the writing, and has had many reviews by several noted DXers and technical gurus as well as Amateurs just entering into the DX scene. My gratitude particularly to W6YOO Harry Hodges, N6KI Dennis Vernacchia, WN6K Paul Dorey, KC6UQH Art McBride, NN6X Paul DeCicco, and N6GY Al Criqui for technical inputs and reading the book for content. Most helpful with the Propagation chapter was K9LA Roland Luetzelschwabm. Many of the Palomar Amateur Radio Club members here in San Diego contributed as well in many repeater conversations. All contributed significantly to the final product. Finally, the most heart-felt thanks to KG6DVD, Harold Locke who undertook the task of the zone maps.

Although I have been a tech writer for 30+ years, I have never reached the level of the Old master, my long time friend, Hal Netten who read every single line, and checked the entire manuscript for grammar, spelling, and punctuation. And last but certainly not least, my XYL Karla Kendrick, who provided a real pileup of encouragement and read through the entire book from the viewpoint of a "civilian" and contributed many innovative suggestions.

Rodney R. Dinkins
Oceanside, CA
E-Mail address is on my web page   http://www.ac6v.com

DISCLAIMER AND ALL RIGHTS RESERVED

CONTENTS

Chapter 3. Operating Aids

Chapter 4. Propagation

Chapter 5. Working HF DX

AC6V DX Reference Guide
Front Matter

Chapter 6. Working 6M DX

PARAGRAPH & TOPIC	PARAGRAPH & TOPIC
6-2. The Big Gun myth	6-10. Worked All States
6-3. Finding DX	6-11. Worked 100 Grids Squares (VUCC)
6-4. Non-Contest DXing	6-12. DXCC Awards
6-5. Working Split	6-13. Other Awards
6-6. Working Contests	6-14. 6M Frequency Allotments
6-7. QRP Operation	6-15. USA Recommended Band Plan
6-8. CW Operation	6-16. Calling Frequencies
6-9. DX Packet Clusters and Telnet	6-17. Tuning In A Nutshell
	6-18. Typical 6M FM Band Plan

Chapter 7. DX Secrets – The coveted revelations of the ages

Chapter 8. Contesting

Paragraph & Topic	Paragraph & Topic
8-2. Art Of Contesting	8-6. CW Contesting
8-3. Phone Contesting Equipment	8-7. Logging
8-4. CW Contesting Equipment	8-8. Submitting The Logs
8-5. Phone Contesting	8-9. Check Logs

Chapter 9. QSLing

Paragraph & Topic	Paragraph & Topic
9-2. The QSL Card	9-15. Envelopes
9-3. Logging	9-16. Envelope Sizes
9-4. QSL Card Design	9-17. Return Postage
9-5. Finding QSL Addresses	9-18. Postage Stamps
9-9. QSLing SWL Stations	9-19. Cash
9-7. Sending QSL Cards	9-20. Bank Checks
9-8. Sending QSL's Direct	9-21. IRC's
9-9. QSLing Via A DX Manager	9-22. Monetary Tips For QSL Managers
9-10. Outgoing QSL Bureaus	9-23. Current US Postal International Rates
9-11. Incoming QSL Bureaus	9-24. What Is A Good IRC?
9-12. QSLing To A DX Bureau Direct	9-25. How Much Postage To Send
9-13. QSL Forwarding Services	9-26. QSL Checklist
9-14. QSLing Electronically	

Appendix A1. CW Operating Procedures

Paragraph & Topic	Paragraph & Topic
A1-2. CW Tutorial	A1-5. Iambic Keying
A1-3. Prosigns For Morse Code	A1-6. Adjusting Paddles
A1-4. CW Bandwidth And Spacing	A1-7. CW Abbrev, Q-Signals, And RST

SAFETY FOR DXers

HIGH VOLTAGE AND LOW VOLTAGE. Linear amplifiers, computers, and older vacuum tube equipment have high voltages inside the case. **WARNING: THESE VOLTAGES CAN BE LETHAL.** Be sure you are properly trained to work on equipment with high voltages. Take all necessary safety precautions when working on equipment. Avoid contact with circuit areas with high voltage, do not spill liquids or drop metallic objects into the equipment. Take off all jewelry, watches and bracelets before working on any equipment. Never have one hand on the metal chassis when making voltage measurements (or poking around) with the other hand. Always use safety glasses when soldering or working with machinery. Although modern solid-state equipment is typically powered with 12 Volts DC, getting a ring or watch across 12 Volts high current DC can frazzle the jewelry and the resultant heat can cause severe burns. Take off all jewelry, watches and bracelets before working on any equipment. **If in doubt about any safety items, seek professional help.**

ANTENNAS AND TOWERS

DO NOT COME IN CONTACT WITH A RADIATING ANTENNA, high voltages can exist on an antenna causing severe RF burns. Make sure that no one can key up your transmitter while you are working on the antenna.

DO NOT MOUNT ANTENNAS NEAR POWER LINES. Make absolutely certain that the entire structure will clear any power lines should the mast, tower or antenna fail.

TOWER INSTALLATION and adjustments is best performed with two or more EXPERIENCED people or professional installers. Heights and towers can represent hazardous situations. There are documented cases of hams being killed or losing limbs while working on towers and elevated antenna systems.

GROUNDING AND LIGHTNING PROTECTION. This is a vital part of your ham shack. Imperative that you read PolyPhaser – URL: http://www.polyphaser.com/pdf/PTD1016.pdf

RF SAFETY CHECKS. The FCC has decreed that you must perform an RF Safety check when filing certain FCC forms. This is a main beam power density estimation intended for use as part of a routine evaluation of RF safety compliance with FCC regulations. These rules can be found in the FCC's ET Docket No. 93-62. More information can be found at the ARRL Web's RF Safety page.

COMPUTER SCREENS. If you stare into a computer screen all day, investigate one of the radiation shields, especially for older monitors. Most newer ones have adopted the latest emission standards.

MEDICAL IMPLANTS. If you or someone in your household has medical implants, be sure to consult your doctor before operating any RF emitting equipment. High voltages are also dangerous for some implants.

MEDICAL FACILITIES. DO NOT operate radios transmitters in or near medical facilities, some medical instruments are sensitive to RF.

COMMERCIAL TRANSPORTATION. Before operating RF transmitters or receivers aboard a ship, airplane, bus or any public or commercial transportation, ask the person in charge for permission.

DON'T's FOR DXers -- Although all is fair in love, war, and DX, observing the following list of don'ts will allow the DX station to realize a faster Q-Rate and hopefully work you sooner.

In split operations -- DO NOT transmit on the DX frequency asking where is he listening – tune around (usually up) and you'll know. To be sure, just listen to the DX operator, they will announce it periodically e.g., this is DX11DX listening up 200 to 250, QRZ. Checking the Packet Cluster will also get this information.

DO NOT call if you can't hear the DX station. Sounds dumb but it is surprising how many folks do this, maybe hoping someone will help them "hear" the DX, e.g., L0GUN he is calling you.

DON'T CALL during other people's contacts, at best it slows down the Q-Rate, may get you on the "Black List", besides being downright rude.

When the DX is calling by district, DON'T call out of turn; you slow things down for EVERYONE. A sharp DX station will ignore you anyway. The DX should make it clear what he/she is asking for. Calling for the 4th call area invites any one with a W4 call or anyone that is portable W4 e.g., AC6V/4. Calling for "call letters W4" is not inviting the portables in the W4 area. Gets confusing but you decide.

DO NOT slow down the DX run by asking for QSL info – most DX operators give it periodically – if the DX runs forever without giving QSL info and you haven't seen it on the cluster or QSL routes on the web, then ask. On the packet clusters, you can query by typing in SH/QSL DX11DX and get the info. The DX Summit is good for this too..

DON'T TUNE up on the DX frequency or the split frequency. DO NOT over adjust your transmitter and spatter all over the band.

DO NOT ask, "What is his call". In this age of DX Packet Clusters it is incredible any one would ask. Good DX operators give it out frequently – just listen. Here again check the clusters.

DON'T play Radio Cop (police persons) – typically one poor guy is out of line and 10 cops are telling him everything from good advice to insults. Best not to do this as your rock-crushing signal covers up the DX and creates chaos and just slows things down.

CHAPTER I   INTRODUCTION

1-1. HOW THIS BOOK CAME ABOUT

I recently had a landline from Bob, a new extra class ham, wanting to chat about DXing. He wanted to know all about it, breaking the pileup, split operation, lists, nets, QSLing, prefixes, packet clusters, the internet, zones, et. al. We talked for over an hour and it occurred to me that many of these questions were the same ones that I had asked of my mentors some 30 years ago.

It was a quite an enlightening conversation for both of us. Fundamentals don't change much it seems. But there is a whole new DX ballgame with the advent of the Internet. Worldwide DX spots can cause almost instant pileups. QSL information is readily available on the web. DXpeditions now have web pages and on-line log checks. And there is a whole host of DX tools readily available from websites such as beam headings, propagation reports, logging programs, prefix lists, contest and DXing programs for the Ham Shack computer.

Back 30 years ago, one found DX by tuning the bands and from the published DX bulletins or the infamous one-ringer from a DX buddy in the middle of the night. About 20 years ago, local DXers were using VHF simplex, then VHF repeaters in an organized DX spotting network. Later the DX Packet Clusters made statewide spotting available. Now there are several world-wide DX spotting networks from various countries around the world. If a rare country appears on the HF bands, seldom do they escape the attention of the spotting networks.

Computers and the Internet really have changed the way we operate and are a necessity for the modern DXer. In addition it has fostered a whole new series of data modes, such as PSK31, MFSK16, and others. No doubt, DXpeditions will be sending not only RTTY but the new modes as well.

Some call books are available on line, but by no means all. QRZ.com offers a good lookup for domestic calls and some DX. A good comprehensive DX call sign lookup is Pathfinder at URL: http://www.qsl.net/pathfinder/   With the exception of the DXpeditions, the Radio Amateur's Callbook (RAC) (flying horse) is still the definitive source.

When Bob asked for book recommendations that address the unique and wacky world of DXing, I gave him some of the titles and authors I was aware of. Bob had read some of these but found them outdated or lacking in beginner's information. He strongly urged me to write a book aimed at beginners and with the new DXers coming on the bands in April 2000, I thought sure why not. So here it is. I hope it helps the new DXer and a few old timers as well. For old timers still not on the Internet, there is a whole host of new tools awaiting you.

1-2.   USING THE GUIDE

Depending on your level of expertise, you may want to skip over the first few chapters and get right to the goodies. You can use the table of contents or the extensive index at the back of the book and start in where you want information.

1-3.   WHY DX

The usual answers are "Why not?" "DX IS" "Because it is there!" But it really depends on whom you ask about this age-old question. When some ancient human first threw a stone, chances are that within minutes a rival tried to throw it further, and so it has been ever since. Run faster, lift more, fly higher, communicate further, climb the highest mountain as a song once chanted.

For Driver DXers, it's power and a burning desire to be at the top of the DX chain thus satisfying their aggressive nature. To this DXer, life is a contest and the DXCC honor roll is one of the ultimate achievements. There is an element of the hunter in the driver DXer, the quarry, the chase, the final contact - all enter into the equation. They bully their way through the pileup with power and ruthless pursuit. Drivers abhor the net users, as they are not playing the game properly, shooting fish in a barrel.

For the Analytic DXer, it's an organized scientific approach, plotting propagation charts, analyzing the MUF and various globe paths. This DXer uses Excel and antenna modeling with exotic formulas to calculate totals, times, take-off angles, beam headings, MUF etc., delighting in adding another checkmark. Usually geography experts, they know every spit and cay around the globe. This DXer can tell you when Armenia will be on the air, what band, and the predicted signal strength. The analytic DXer tends to be the first one to find DX and work them.

The Amiable DXer joins every DX club and association in sight, loves to tell DX stories and be "one of the crowd". When not chasing a new one, this DXer loves to ragchew with the DX stations. Tends to be polite in a pileup, but not too polite! Very patient type, will call for hours and finally work them.

The Expressive DXer loves power through people, rises to club president, knows many rare DX operators on a personal basis, and has regular schedules with noted DXers overseas. Shares many characteristics of the Driver.

But there is a little of these personalities in all DXers, the hunt, hurling our signal around the world to exotic places, the knowledge of propagation and MUF, geography, checking them off, telling DX stories, being admired by our peers, being recognized by noted DXers. Besides the rational reasons for DXing, underneath there is a mystique, an enigma about it all.
The great Hugh Cassidy, WA6AUD expressed it this way:

Then shook the hills with thunder riven,
Then rushed DXers to battle driven.
And louder than the bolts of heaven,
"I QSL, you're five by seven..."
Not for thee the joys of family,
Nor the siren call of sex.
Your vocation you've decided
Is chasing the rare DX.
Thus is the law of DXing
That only the strong shall thrive
That surely the weak shall perish
And only the Deserving survive!
Hugh Cassidy – WA6AUD

Why don't ya come on over and work me
sometime – Mae East

1-4. THE HF BANDS

Amateur Radio operators are blessed with a wide range of HF frequencies from 160 meters to 10 meters. Depending on solar conditions, world wide contacts in many modes can be made using a modest station and antennas. Activities include general conversations, contesting, DX contacts, digital modes, emergency service, traffic handling, low power operation, mobile operation, and many more exciting world wide activities. During the peak of the solar cycle, the bands can be packed with signals from all over the world. Even during the lows of the solar cycle, Amateurs can still make world wide contacts on the lower frequency bands. A charter of Amateur radio is the continuation and extension of the amateur's unique ability to enhance international goodwill, and indeed many friends can be made world wide via the HF bands. It is not uncommon to travel abroad and be hosted by Amateurs that one contacted on the HF bands.

1-5. THE SIX METER BAND

The 50 MHz Amateur Radio band is considered a VHF band and is perhaps the most intriguing of our bands. It can exhibit several types of normal propagation and at times – very strange propagation modes. Since the band was opened up way back when, there have been many hams on the band that have 30 plus years experience on the magic band and many of these work this band only. This book is not the definitive guide to six meters, but rather is intended to help beginners into six meter DXing.

Back when television was in its infancy, I was watching a Cleveland Ohio TV station 50 miles away on channel 2 and all of a sudden, bars started scrolling all over the picture and within moments the Cleveland station was replaced by a TV station from Texas! The phenomenon lasted for a few minutes and then the station from Cleveland reappeared!

I asked my Ham elmer what was going on and he explained the exotic term sporadic E propagation. Although it would be several decades before I obtained my Ham ticket, the experience remained a mysterious occurrence in my memory banks.

In November of 1999, I purchased a Kenwood TS-60 Six Meter rig, put up a Ringo half wave vertical and set about to test the new installation. The first check would be to access the local six meter repeater. After fiddling with offsets, splits, and PL, NOTHING. So I tuned the SSB portion and the beacons NOTHING.

Help was in order, so I called my neighbor who had six meters and he said he would meet me on 50.125 MHz - the National SSB Calling frequency after he finished his phone call. As I tuned in to the calling frequency, I heard N0KE/KH6 calling CQ and receiving no answers. I called, he answered, and I had my first 6M QSO. Little did I know what had just happened, I had just worked Hawaii - one of most difficult US states for Worked All States. Later I would be informed the local six meter repeater was off the air.

Later I would realize that we were near the top of sunspot cycle 23 and the propagation from southern California to Hawaii was via F layer skip. Soon I was working back to New England and within a few months had worked every state on the eastern seaboard from Maine to Florida – all on F layer skip. On one occasion, signals came in from the Caribbean and Cuba. Later Alaska was snagged during an F-Layer skip to the north. Well now, we had many of the distant states, but little else. Hey this was like 10 meters – but I was due for a shock when the solar cycle declined and F layer skip went away.

This was all very satisfying but where were the other states, I wondered. A local guru told me that they would come during sporadic E season in the summer, and sure enough, propagation to the western portion of the US became evident that summer.

Working all US states on six meters was a goal I wanted to achieve. But again how does one work the Midwestern US states from California?

The answer came at the ARRL Southwest Convention where an amazing forum on six meters was conducted. In this forum, a 30 year veteran of the six meter band discussed propagation, equipment, the esoteric modes such as meter scatter and EME. While discussing sporadic E propagation, the answer to working all US states was revealed – double hop sporadic E.

Even so, when the double hop openings occurred, the quest for WAS on six meters would be a long haul. To date 48 of the 50 states have been worked, 6 DXCC Countries and 198 Grid Squares (VUCC) Working DXCC on six meters could be a life time endeavor but many have done it.

1-6. THE JOY OF DXING

Early morning before sunup, the family is asleep, and Old Shep is snuggled up against your feet. The quasi-military looking transceiver glows a fluorescent blue and everything you've assembled yourself is tuned finer that a gnat's eyebrow. All is well, the coffee cup steams with fresh brew and after checking the propagation, you swing the beam to Asia and tune the 20 meter CW band with a practiced ear.

There is a strong but fluttery carrier, then a QRL DE BS7AF (frequency busy?), good grief – may be the reef, Scarborough Reef in the South China Sea!! A rare find in so much as the last DXpedition was shot at by some strange looking characters ashore. Could be a pirate. Well work them now, worry later. Old Shep yawns as you call back at a little lower speed than the BS7. Son of a gun, no pileup and you have a nice chat assuring the BS7 operator that they have excellent propagation into San Diego and a fair idea that they are legit, QSL manager and all. Beats mowing the lawn before going off to work. It will be good day.

THE SKED
Hello Pitcairn Island
Could you have
VP6DXX meet me
on 21.298 Mhz?

1-7. THE DXER AND THE CIVILIAN

If you dare to discuss DXing with a non-ham, it will probably go like this:

Civilian – You look like something the cat drug in – good party huh?

DXer – Nope I was up til 3 AM, man what a rush – I worked the Kingman Reef DXpedition last night on Ham Radio.

Civilian – whatta you mean by worked?

DXer - Well there was a terrific pileup on 20 Meters and after 2 hours, I broke through.

Civilian – (interrupting) pileup, 20 Meters, I don't get it, what is a DXpedition anyway?

DXer - It's an organized radio expedition to a rare country or island, the purpose is to put the rare location on the air and contact as many Hams as possible. Some DXpeditioners have spent as much as $400,000 to get to an uninhabited island.

Civilian (Looking at you strangely) – You got to be kidding!

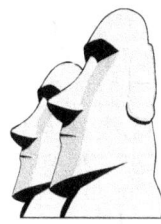  

> Oh Cripes !!
> Not Another
> DXpedition

From the civilian's perspective – why on God's earth would a rational human being shout (key) for hours into a mob of ham stations to finally get a "ur 59(9) QRZed", perhaps not even knowing the DX's name. Well the civilians and non-DXing hams just don't understand, do they? As Cass alluded "You have to be a believer, a true blue DXer in order to understand". Tally Ho! It's the hunt, old chap!

On the serious side, one of the charters of Amateur Radio is Emergency Communications and when an earthquake hits half way around the world – DXers are an invaluable asset to aid in international disasters. The charters of the Amateur Radio Service include emergency communications and fostering international good will. We DXers are making connections, the human touch, perhaps in some small way making peace and improving the human condition.

Some folks race cars, others strive for the best bowling average, DXers engage in Radio Sport, Contests and DXing, hurling our messages around the world. Sir Edmund Hillary, when asked why he was going to climb Mt. Everest, either was being very flippant or very profound, when he replied, BECAUSE IT'S THERE! And so brethren is Heard Island!

1-8. MORE JOY OF DXING

It's late at night, the family is asleep, the lights are down low, the radio dials glow reassuringly. You're armed with an 807 (ham slang for a beer), a java script DX page on the Internet, the SFI is 250, and 20 Meters is open worldwide. You are stalking EK6XXX, you know from the OH2BU packet cluster, she comes on at 0400Z at about 14.206 on Monday nights. Maybe her OM attends the Yerevan poker game on Mondays, who knows? You check the MUF, adjust the beam to the proper bearing, fire up the linear and you are ready to go at 0330Z.

0342Z. Calling CQ CQ CQ EK6XXX calling CQ and bye. SON OF A GUN – there she is, early! You're ready, she called, you calmly and professionally reply: EK6XXX here is WA6WTO, good morning Anna. With a little surprise that you know her name, she returns to you despite a few big guns also calling.

It's a pleasant QSO as most contacts are when you catch a DX station first calling CQ. You both sign off and the band explodes – the packet clusters all over the world have spotted EK6XXX.

Well now, you have done your homework, it worked just as planned and that is the *joy of DXing*. Time for an 807 to celebrate by golly, as you mosey on down to the fridge shouting and waking up the whole family "I got em – Armenia Armenia on 20 Meters". The XYL yawns and says "I worked an EK in the contest last week" and the harmonics say "Geez Dad – couldn't you send Charlie Whiskey"?

Its 10:00 AM Tuesday and you are at the salt mine slaving away. At precisely 10:00 AM, you tell the boss that the washing machine is dumping water all over the house. After a lifted eyebrow and a curt remark of "that happened last year, didn't it, you're excused". Now for a mad dash to the shack to work the DXpedition that just came on from Charcoal Island in the Barbecues. You get them with no problem as every other DXer is at work or is retired and already worked them, and you're back to work in a half hour. Sixteen other hams in the company also had broken washing machines, fortunately they all work in different departments! *The joy of DX.*

It is contest weekend and you're using the event to get a new CW DX band country. S92UU is running an enormous pile, split operation. You don't transmit just yet but spend a few minutes using your transmit track control and track how the operator is working the pileup. The DX op tunes up catching an in-the-clear call, then when the 5 kHz split is reached, goes back to 2 kHz up. You're ready, the 5 kHz point is reached, you fire away at the 2 kHz point and it's in the bag. *The joy of DX.*

---

[1] DXCC is ARRL Copyrighted. Permission to use here has been granted.

Kingman Reef is on 15 meters, good grief it's the reef – for an all time new one. You call periodically for hours, no luck. The band begins to drop out, no W1's now. Time passes, only W6's are being worked, but not you. Agh, the power line noise is getting heavy. The Kingman signal drops into the noise – Good Golly Miss Molly they are still at it. You tweak up the DSP unit, clamp on the titanium headphones, and strain for the syllables through the noise. All of a sudden, Kingman calls QRZed and no one answers. You call WA6WTO, they answer, and you are in like Flynn! Ain't no power line noise on Kingman Reef!!!

You need a change so you adjust the rig for 5 Watts out. 15 Meters is hot. YB0XXX is S-9. You scream WA6WTO QRP, no dice. 10 minutes later, you are receiving S-9 +10dB, no cigar. Another 10 minutes pass and the YB0 is 20 dB over – you call and roar right through the W1's and crack the pileup with 5 Watts. Ah yes, propagation is!

1-9. DX IS FOR ALL

DXing and Amateur Radio are open to all: he, she, them, us, youths, seniors, handicapped, all have made their mark. The YL contingent has contests, nets, clubs, and seem to break a pileup quicker than any one else (high frequency notes are!). With the stamina of youth, youngsters do well in contests and bring in a fresh exuberance to the sport. The ARRL sponsors a Scouting Contest and other activities and awards.

1-10. DX CLUBS AND FOUNDATIONS

DX Clubs are located in many cities around the world. They offer camaraderie, elmering, DX and DXpedition presentations, club contests, a chance to be part of a DXpedition, and a valuable source for DX information. Also a great place to tell DX lies. Many of the clubs operate DX packet clusters for spotting DX heard in your area. DX Clubs have ARRL appointed QSL card checkers that can check most cards for DXCC credit.
W6YOO is the checker for the San Diego area. There are too many DX Clubs throughout the world to list here. They can however be found on the Internet   http://ac6v.com/clubs.htm

There are several DX Foundations throughout the world that assist worthwhile Amateur Radio and scientific projects with funding and equipment. A typical example is the Northern California DX Foundation. (International Beacon Program to be discussed later.) These can be found on the internet at URL: http://ac6v.com/clubs.htm.

1-11. NATIONAL AMATEUR RADIO LEAGUES

Most countries have an Amateur Radio League or Association that serves in the interests of the Amateurs in their country. Examples are the USA ARRL, Canada's RAC, UK's RSGB, France's REF. Many of these are on the Internet   http://ac6v.com/clubs.htm

1-12. AWARDS

Probably the biggest award for the DXer is the satisfaction of building an Amateur Radio Station and hurling your voice around the globe. Somehow I came up with a biblical sounding theme back in the early 80's that - for me - says it all:

Hark! I Have Hurled My Words To The Far Reaches Of The Earth! What King Of Old Could Do Thus? --- © AC6V

The DX Century Club is the premier operating award of Amateur Radio. The basic certificate (which can be obtained in several categories -- mixed modes, phone, CW, RTTY, Satellite, 160, 80, 40, 20, 15, 10, 6 and 2 meters) is awarded for working and confirming at least 100 entities on the ARRL DXCC List. Endorsements are available in specific increments beyond the 100-entity level, culminating in the coveted DXCC Honor Roll or Top Of The Honor Roll, for those near to or have worked them all! Further DX Century Club information is available on the Internet at http://www.arrl.org/awards/dxcc/. A **5BDXCC award** is also available and endorsable for the 160, 17, 12, 6 and 2-meter bands. Several versions of WAS – worked all US States are also sought after.

CQ Magazine DX Awards. CQ magazine sponsors several awards including the CQ DX Awards Program, the CQ Prefix award - WPX, and the CQ Worked All Zones – WAZ.

There are thousands of awards available from all over the world, and wallpaper chasers are well advised to purchase the **K1BV DX Awards Directory** By Ted Melinosky covering over 2,780 different awards, certificates and trophies from 122 DXCC countries. Also folks get into Islands-On-The-Air (IOTA), US County, and Ten/Ten International awards.

1-13.   BRIEF HISTORY OF AMATUR DXING

1920 to 1980's Don Wallace W6AM. DX Hall of Fame and early pioneer of Amateur radio. Don has probably done more to promote DX operation and encourage new operators than any other individual. Famous for his antenna farms in Rolling Hills on the Palos Verdes peninsula.

1920 The Radio Amateurs Callbook (RAC, Flying Horse) is published. International QSL bureaus are established.

1923, November 27, the impossible happened. Leon Deloy (8AB), of Nice, France worked (on 110 m CW) USA stations: Fred H. Schnell (1MO, Connecticut) and John L. Reinartz (1QP/1XAL, after - W3RB). Four thousand miles - DX For Sure.

1924, Oct 18 A station in England G2SZ Cecil Goyder worked a New Zealand station Z4AA Frank Bell, a distance of almost 12,000 miles.

1926, Brandon Wentworth, 6OI, achieved confirmation for working all of the continents.

1927-1982 KV4AA Dick Spenceley in the U.S. Virgin Islands provides thousands of contacts over the years. He was inducted into the CQ DX Hall of Fame in March, 1969.

1936 - 56 Mcs - G5BY was the first European to span the Atlantic on 56MHz when his signals were heard by W2HXD.

1937 The ARRL introduces the DXCC Program. Discontinued during WWII and started all over again after the war.

1947 - The DXCC country count for this year was 257. Gatti-Hallicrafters Africa DXpedition Nine-Month Tour. The Quarter Century Wireless Association was organized. The Old Old Timers Club was founded in 1947.

1955 - 1963 Danny Weil DXpeditions - Starts from England and in 8 years gives contacts from 30 different countries.

1960 - 1970 Gus Browning (W4BPD) The first DXer elected to the DX Hall of Fame. Operated from over 100 countries.

1961 - Present OH2BH Martti Laine one of the most accomplished DXers of our time.

1962 - 1967 Don Miller W9WNV Many DXpeditions.

Over the years, many of the well known DXers include: K2GL, KH6IJ, G3FXB, OH2BH, W8IMZ, W3GRF, W3GM, W4BPD, W1WY, W2PV, W3AU, K3ZO, W9WNV, W4KFC, W7RM, W1BIH, PY5EG, W6QD, N6TJ, S50A, N6AA, K1EA, OH2MM, K4VX, K3EST, W6RR, ON4UN, LU8DQ, K1AR, N4MM, VP2ML, W6AM, KV4AA, W1FH, W6RGG, W6RJ, W1CW, W6ISQ, W6OAT, W6KG, W6QL. Many of these were inducted into the CQ DX Hall. For more – see URL: http://ac6v.com/history.htm Top Contesters include K7JA, KH6IJ, and W0UA, W7RM, K6KII, KG6AAY, N5TJ, N6TR, N6AA, W0UA, N6IG, OK1RR

I tell you, the guy down there is a nut
–sends "seek you" into my TV set

CHAPTER II   DXING EQUIPMENT

2-1. CHAPTER CONTENTS

This chapter contains information on equipment for DXing with emphasis on the selection and characteristics of the equipment. The table below lists the chapter topics and paragraph numbers where they can be found.

PARAGRAPH & TOPIC	PARAGRAPH & TOPIC
2-2. Transceivers	2-30. Six Meter Antennas
2-3. Transceiver Specifications	2-31. 6M Antenna Height
2-4. Sensitivity	2-32. Homebrew Antennas
2-5. Dynamic range	2-33. Low Band Antennas
2-6. Selectivity	2-34. 40 And 30 Meter Antennas
2-7. Superheterodyne Receivers	2-35. Stealth Antennas
2-8. Frequency Stability and Accuracy	2-36. Antenna Gain
2-9. Cross -Modulation	2-37. Towers & Rotors
2-10. Gain Compression	2-38. Bearing And Distance
2-11. Intermodulation Distortion	2-39. Grounding And Common Mode
2-12. Phase Noise	2-40. RFI Safety And RFI To The Neighbors
2-13. S-Units	2-41. Lightning Protection And Grounding
2-14. Decibels	2-42. Computer Equipment
2-15. Transmitter Specifications	2-43. Coax Cable & Cable Management
2-16. Duty Cycle	2-44. Voltage Standing Wave Ratio
2-17. Whizbang DSP Radios	2-45. Antenna Tuners
2-18. Transceiver Programmability	2-46. Microphones
2-19. Transceiver Operating Controls	2-47. Headphones And Speakers
2-20. Your Sideband Audio	2-48. ESP Hearing
2-21. Modifications	2-49. Keys And Keyers
2-22. Linear Amplifiers	2-50. DSP Units
2-23. Selecting Power Amplifiers	2-51. Antenna Analyzers
2-24. Tubes VS. FETS	2-52. SWR Meters
2-25. HFAntennas	2-53. Phone Patches
2-26. HF Antenna Height	2-54. Scratch Pads
2-27. Yagi Beams	2-55. Monitor Scopes
2-28. HF Vertical Antennas	2-56. Foot Switches
2-29. Cubical Quads	2-57. Other Goodies

2-2. TRANSCEIVERS

Modern transceivers are a marvel of design and features. In general, the higher the price, the more features and performance can be expected.

The best rig for contesting may not be the best one for DXing, CW operation, or all around rag chewing. Contesters or CW ops will tell you of their favorite rig as well as DXers their favorite. Solid-state rigs have the advantages of no-tune, quick band changing, general coverage (SWL listening), memories, scanning, high accuracy as compared to older vacuum tube rigs that used analog tuning. Modern rigs also offer noise reduction, tailored audio, and lots more. But they can be finicky about heat and high SWR – reducing power when out of the SWR rating.

For contesting, front end characteristics such as dynamic range and selectivity and filtering are of paramount importance in the big-signal, jammed band environment, where DXers might be more concerned with sensitivity, signal to noise characteristics and DSP for discerning weak signals. For some favorite transceivers of contesters and DXers – see Contest Archives http://www.contesting.com/FAQ/

HF Rigs. For casual HF work, the lower price radios may be a good choice, but if you intend to get into big-time contesting, serious DXing, sustained operation such as RTTY or contesting, then the more pricey radios or a used high-end radio may be your choice. Here is a consensus of what DXers/Contesters use for DXing and contesting radios. Not necessarily the best – but what they bought.

Yaesu FT-1000MP, FT-1000D, FT-990, FT-1000MP-MKV, FT-920, FT-1000, Kenwood TS-850, TS-940, TS-950's, TS-570D, TS-870, TS-830, TS-440, TS-450, ICOM IC-706 MKII, IC-756 PRO, IC-735, IC-751A, IC-706, IC-756, IC-738, IC-781, IC 775DSP, Ten-Tec Omni Series – particularly for CW. Also the Elecraft transceivers have FB specifications.

Contesters choices are Yaesu FT-1000MP, FT-1000MP, MKV, FT-1000D, Kenwood TS-850, TS-940, ICOM IC-756PRO, IC-775DSP, IC-765, Ten-Tec OMNI VI+

Without endorsement from the author, best read the rig reviews and comparisons – Elecraft http://www.elecraft.com/ eHam Reviews http://www.eham.net/reviews/ Yahoo Groups http://groups.yahoo.com/search?query=amateur+radio INRAD at http://www.qth.com/INRAD/

There are many others suitable for DXing, but more noted DXers use the above more than any others. There are others that when modified, are suitable for serious DXing and Contesting. INRAD and Sherwood offer improved filters for a wide variety of transceivers that transform the ordinary transceiver into a top notch machine. One thing for DXers/Contesters to avoid is the stripped down entry radios, a used 3 to 5 year old, medium-to high-end rig would be a much better choice. For example, one "entry level" rig has no RF Gain control which you will find indispensable for DXing. Another "entry level" radio just plain shuts down under sustained RTTY operation. Another has no Monitor function and you do want to hear your audio from time to time during a contest or when changing microphones, or adjusting transmit audio. One manufacturer includes a noise blanker, but with no level adjust.

Six Meter Rigs. Today, it's easier than ever to get on six meters. Many of the newer HF rigs come with six-meter capability built in. There also are transverters, such as the ones from Ten-Tec that will configure your HF rig for six meters.

Some single-band rigs are available such as the Ranger six meter rigs. If you're interested in DX, avoid the FM-only six-meter rigs and get one capable of FM, CW and SSB operation. Unlike the HF bands, high power is not necessary for many propagation modes, such as F2 skip and sporadic E. Many 100 Watt radios are available and are adequate for the beginner.

The multiband transceivers come in two flavors, base and mobile/portable. The base rigs are typically higher priced, but offer higher performance and advanced operating features. Examples are Icom IC-746PRO, IC-756PROII, Kenwood TS-2000, TS-570S, and Yaesu FTDX-9000, FT-847. The compact rigs are adaptable to portable and mobile use, examples are Alinco DX-70TH, Icom 703, 706 Mark II G, Yaesu FT-817ND, FT-857D, FT-897, and Kenwood TS-480SAT. The Universal Radio store in Ohio has a well organized list of the current multi-band transceivers. URL: http://www.universal-radio.com/catalog/hamhf.html

Be aware that the compact rigs might not have specifications as good as the base rigs, so study the specs and reviews carefully before deciding on a new rig. Weaknesses such as intermod, dynamic range, birdies, and phase noise may show up in the lower price radios.

2-3. TRANSCEIVER SPECIFICATIONS

Following are discussions of transceiver characteristics written for the DX operator as a guide to selecting and evaluating equipment. In-depth design references can be found in engineering textbooks or the ARRL Amateur Radio Handbook. Reading QST reviews of transceiver performance will help select the radio you can afford. For an excellent engineering book, see "Communications Receivers by Rohde, Whitaker, and Bucher" ISBN 0-07-053608-2, a McGraw-Hill Publication".

Communications receiver design is inherently a process of compromise. Noise limits the minimum signal that a receiver is capable of processing while the maximum signal capability is limited by distortions caused by non-linearities. These limits define the dynamic range of the receiver system. Large interfering signals can also affect the reception of small signals. A kind of catch 22 is when designing for optimal noise characteristics; it typically yields less than optimal large signal performance. Conversely, the best large signal performance suffers from higher noise degradation, which in turn limits the weak signal reception. Let's take a look at transceiver characteristics as they affect you the DXer.

2-4. SENSITIVITY

Sensitivity is the capability of a receiver circuit to detect weak signals and the major factor in receiver sensitivity is the signal-to-noise considerations. Due to the resistance and temperature of various components, receiver noise is inherent in any receiver. In lab measurements, the amount of signal input required to produce a signal to noise ratio of 10 dB is generally used to specify the receiver sensitivity specification.

In modern transceivers, this is typically a few tenths of a microvolt to a few microvolts depending on the input frequency, mode, and bandwidth of the receiver. Another figure of merit used in determining sensitivity is the noise floor, which is another way to express the receiver noise. Typical values are −130 dBm to −140 dBm depending on the mode, filtering and preamplifiers used. The higher dBm numbers are better.

However atmospheric and man-made noise enters into the real environment so that the minimum required sensitivity is something quite different than the lab measurement. On the lower bands the noise can be quite heavy; an S-Meter reading of S7 of noise is not uncommon. So even though the receiver has excellent sensitivity, it is unusable in the presence of atmospheric and man-made noise. Using well-designed DSP units and low noise antennas (beverages and loops) are necessary, particularly on the lower bands. Increased sensitivity is gained at the expense of dynamic range, the latter being of greater importance in today's crowed bands and the noise, both atmospheric and person-made noise.

Discerning weak signals generally requires a signal to noise ratio of 10 dB. The noise is a combination of atmospheric noise, receiver noise and circuit design. Note that many CW operators can typically copy code at a signal to noise ratio of almost 0 dB perhaps accounting for the superiority of CW over phone under minimal signal conditions.

The six meter band does exhibit some atmospheric noise as well as person-made noise. These can override the front-end noise figure on about all the rigs on the market today unless you have a very quiet location. Most rigs include a preamplifier for additional gain for weak signal work.

2-5. DYNAMIC RANGE

Dynamic range is expressed in dB where the lower limit is the smallest discernable signal (receiver noise floor) and the upper limit is the point where intermodulation products become noticeable. It is an important specification as it gives a figure of merit for evaluating the strong signal handling characteristics of a receiver. Values of 103 dB are typical. The use of front-end attenuation and an AIP circuit can help reduce the effects of intermodulation. The receiver noise floor can be affected by receivers using synthesized tuning schemes. This has improved considerably with the new transceiver designs.

Another measurement of this is blocking dynamic range, which is the difference between the receiver's noise floor and a signal which causes one dB of receiver gain compression (i.e., non-linearity). This marks the point where receiver de-sensitivity begins. Very important when trying to pull out a weak DX signal with strong stations nearby.

Intermodulation distortion dynamic range is the difference between the receiver's noise floor and two equal received signals which produce distortion three dB above the noise floor. This marks the point where undesirable spurious signals are generated in the receiver. There are more specs regarding dynamic range – see http://www.elecraft.com/ Also see Sherwood Engineering at http://www.sherweng.com/table.html

Nothing is more disconcerting than to be in a contest or pileup environment and have a receiver that can't handle the huge signals and the weak ones adequately, this coupled with poor selectivity is a lose-lose proposition.

2-6. SELECTIVITY

Selectivity is the receiver characteristic that filters undesired signals from the desired one. Some of the circuits involved with the receiver's selectivity are fixed such as the IF filters.

Other selectivity circuits can be controlled by front panel controls. Various methods are used including SSB Bandwidth, IF Shift, Passband Tuning, Continuously Variable IF Bandwidth, CW/Digital Bandwidth, Notch Filter, and on the newer rigs, digital filtering.

Filters for SSB Bandwidth are generally crystal filters, mechanical filters or variable bandwidth tuning (VBT). Typical bandwidths for good SSB reception are in the order of 2.7 kHz and 2.4 kHz at the 6 dB points. Filters as narrow as 1.8 kHz can be used for SSB but intelligibility suffers and tuning can be critical. VBT allows narrowing the bandwidth continuously.

IF shift and passband tuning schemes allow the passband to be shifted without retuning the receiver. For IF shift, the bandwidth remains constant while the position of the passband can be altered to minimize interference from close by stations. However, as you move the control to eliminate interference on one side of the signal, you may uncover another signal on the other side as the passband is not narrowed in this scheme, only moved. Passband tuning, as implemented on the some ICOMs, moves only one edge of the passband

The new digital receivers allow setting of both the low frequency and high frequency roll off points of the filtering. Settings of 50+ or more roll-off points are available in the DSP radios. In some DSP radios, selectable sharp and soft filter skirts for SSB or CW are available. These alter the audio response curves for various listening conditions.

An important parameter for filters is its shape factor, stated a different way – skirt selectivity. This is expressed as the ratio between the bandwidth at 60 dB and that at 6 dB. Obviously the further down the skirt an adjacent signal is, the less the interference. Shape factors of 1.5 or better are achievable with quality components. Several after market vendors (INRAD and Sherwood) offer modification kits to improve the receiver selectivity. Minimum ripple and symmetry are also important. Crystal filters for CW are typically 500 Hz, 270 Hz, and 250 Hz. The newer DSP radios offer razor-sharp variable CW filtering, some without the use of crystals.

Notch filters are effective against carriers that exist in the passband along with the desired signal. Depending where the notch filter is installed, notching as deep as 60 dB can be obtained on the non-DSP radios.
On the newer DSP rigs, notching is so effective that it virtually eliminates the carrier interference when some one "tunes up on frequency". The notch filters are also effective against heterodynes and birdies.

2-7. SUPERHETRODYNE RECEIVERS

Superheterodyne receivers reduce the incoming signal frequency by mixing in a signal from a local oscillator to produce an intermediate frequency (IF). Superhets have better performance because the components can be optimized for a single intermediate frequency, and take advantage of selectivity designs. Along with front end selectivity, the choice of 1st IF frequency determines rejection for signals at the image frequency. For double conversion superhets, the 1st IF selectivity, along with the 2nd IF selectivity determines the adjacent channel selectivity. All of the above are significant in determining adjacent channel blocking. More gain towards the front end is worse, and more selectivity towards the front end is better, in both cases, costs are reduced.

Triple conversion is not necessarily the best for any of the above, but it does add design flexibility. It can allow more practical (in price or availability) choices for bandpass tuning, less complex transceive functionality with a matching transmitter, common modules for multiple equipment design implementations, and perhaps allow lower cost components. (e.g., LC circuits as opposed to more expensive channel filters.)

2-8. FREQUENCY STABILITY AND ACCURACY

Modern day receivers use a synthesized tuning scheme where the accuracy and stability are determined by a reference source, usually a crystal. Today, with the state of the art design, accuracy and stability of stock transceivers are sufficient for Amateur radio work, although the master oscillator should be calibrated periodically as per the instruction manuals. Increased accuracy and stability can be achieved with a TCXO (temperature controlled crystal oscillator) which many manufacturers offer as an option.

Typical standard crystal units are stable to within +/- 10×10^{-6}, a TCXO can achieve stabilities of +/- 0.5×10^{-6}. Since crystals drift off frequency over time, the transceiver will require periodic calibration. Refer to the manual for calibration instructions and adjustment locations.

Accuracy is determined by the crystal reference and its calibration, resolution is simply the least significant digit on the display. Keep in mind if the crystal has drifted, you can be transmitting or receiving on an actual frequency other than what the display reads. But crystal drifts are generally small over a fairly long period of time. Calibrate periodically.

Since most of the frequency tuning on the Amateur bands is to adjust for clear speech on SSB or adjust for pitch or zero beat on CW, an exact frequency setting is usually not critical. The standard crystal supplied with modern transceivers is more than adequate to determine band edges or to tune to a specific frequency. In fact, today's rigs are accurate enough so that DX Packet Spots will put you very close to the DX frequency. But PSK31 tuning might require a TCXO or more frequent calibration since PSK is very narrow bandwidth.

A TCXO might prove helpful for beacon stations or net control stations desiring the utmost in performance, or in severe environmental conditions, such as Arctic expeditions, or desert locations.

Older rigs use LC circuits (analog) to tune the receiver. Since these rigs are not crystal controlled, drifting is not unusual. Either inboard or outboard crystal oscillators are used to calibrate the main dial. These radios might present a problem when tuning PSK31.

2-9. CROSS-MODULATION

Cross-modulation is intermodulation caused by the modulation of the carrier of a desired signal by an undesired signal. It is independent of the desired signal and proportional to the square of the undesired signal. Using the input attenuator or AIP (Kenwood) on a transceiver can help reduce the effects of cross-modulation.

2-10. GAIN COMPRESSION

This nasty occurs when a strong signal drives a receiver stage into saturation and can be discerned with a decrease in background noise. For multi-contest environments or a nearby Ham transmitters, the use of outboard filters help solve the problem.

2-11. INTERMODULATION DISTORTION

IMD or intermodulation distortion is caused by two strong signals that drive the receiver front-end circuits (first RF amplifiers or preamplifiers) beyond their linear range, thus producing spurious signals. The most common IMD is third-order products. The measurements for determining IMD get quite involved so you may want to refer to engineering texts for more information. Suffice it to say that when signals get into the S9 +20 dB plus range, IMD may occur and can be reduced by the receiver's attenuators and AIP function.

2-12. PHASE NOISE

Phase noise is a shortcoming of using synthesized tuning schemes. The modern transceiver employs synthesized tuning that allows for very accurate frequency settings and excellent stability. In addition, the design makes it feasible to make the receiver cover all frequencies in the HF spectrum thus giving a bonus for shortwave listening. Unfortunately, synthesizers can add noise into the receiving system and is a consideration when evaluating a receiver. The newer radios are considerably improved in this regard.

2-13. S - UNITS

"You're hitting my S-Meter at S7", reports the new Ham, but this means little to the operator on the other end, as we shall see in a moment. The proper way to give reports is with the RST system in the Appendices, and S7 ain't the same as 57. At best, S-Meters are useful for relative comparisons. In fact, there is no standard for S-Meters. There is an old suggested standard whereby S9 is calibrated at 50 microvolts RF input and then 6 dB per S unit down from that reference point.

Other sources say 5 dB per S-unit and this seems to be more common than 6dB/S-Unit. Engineers and users have checked a variety of rigs over the years and found a wide variance among radio brands, variance among modes with the same rig, and variance among bands on the same rig. But here are some typical results that can be used for a rough idea what an S-Meter is reading and the corresponding microvolt level.

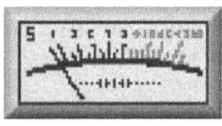

MicroVolts	S-Units	MicroVolts	S-Units
0.2 uV	S1	6.3 uV	S6
0.4 uV	S2	12.5 uV	S7
0.8 uV	S3	25.0 uV	S8
1.6 uV	S4	50.0 uV	S9
3.2 uV	S5	158.0 uV	S9+10dB

There are many more specifications pertinent to the DX operator, but these are the main ones to consider when selecting a transceiver. You may want to refer to engineering texts or the ARRL Amateur Radio Handbook. Also QST equipment reviews are helpful. A good source of transceiver performance can be found on the Internet. Also one can read what users think of the various radios at URL: http://www.eham.net/reviews/

These can be somewhat subjective, but will give you an idea of some things to look for when evaluating a radio. A good example is that for some radios, early models had problems and required modification, so if you are buying a used radio, this web site is a must see.

Transmitter specifications are a whole different subject, but since we don't homebrew much these days and the equipment must be type accepted, all modern rigs are very good in this regard. The QST reviews will show the transmitter performance characteristics.

2-14. DECIBELS

The decibel (1/10 Bel) (dB) represents a logarithmic ratio between two quantities and is unitless. It is a more representative way of expressing gains and losses of electronic elements. Doubling power is approximately 3 dB. An easy way to remember dB is to consider doubling of a typical transceiver output. From 100 W to 200 W is 3 dB; 200W to 400W is 3 more dB or 6 dB above the 100 w level (6dB is roughly one S-Unit on a receiver). Continuing 400 W to 800 W is 3 dB and 800 W to 1600 W is 3 more dB. Thus going from 100 W to a 1600 W is 12 dB or 2 S-Units on a receiver. In power ratios, dB=10 log (P1/P2) so 200W/100W = 2. The log of 2 is 0.301. Then 10 times = 3.01 dB.

If the ratio is referred to a specific quantity, this is indicated by a suffix (dBm is referenced against 1 mW and dBV is 1 Volt). Since power is proportional to voltage squared, the ratio of voltages or currents across a constant impedance is given by dB=20 log (V1/V2) or 20 log (I1/I2). Simple rule of thumb: When working with power, 3 dB is twice, 10 dB is 10 times. When working with voltage or current, 6 dB is twice, 20 dB is 10 times. Minus dB readings are similar, 3 dB losses is half the power lost over the device.

2-15. TRANSMITTER SPECIFICATIONS

When Amateurs built their own equipment, meeting the FCC requirements for transmitters was a real challenge; today we rely on the manufacturers to produce equipment that meets specifications. QST magazine performs extensive testing on the transmitter section of a transceiver, for members only, see URL: http://www.arrl.org/members-only/prodrev/

Major areas of transmitter testing are:

1. Output Power
2. Transmit Frequency Range
3. CW Transmit Tests
4. Spectral Purity
5. Two-Tone Transmit IMD

6. SSB Carrier and Unwanted Sideband Suppression
7. CW Keying
8. PTT to SSB/FM RF Output
9. Transmit/Receive Turn-Around
10. Keyer Speed and Sidetone Frequency
11. Composite Noise

For an extensive discussion of how these tests are performed – see URL:
http://www.arrl.org/members-only/prodrev/testproc.pdf

2-16. DUTY CYCLE

For the transmitter section of a solid-state transceiver, there are heavyweights and lightweights. The lightweights have poor performance on sustained carrier operation like RTTY or Amplitude Modulation. Many will protest and throttle down well below the 100 Watt level. Some will just shut down when the unit gets too hot.

On the other hand, the heavyweights have fans, and large heat sinks for the finals and will run all day on RTTY without a whimper. Sustained contesting can separate the heavy and lightweights. Contesters are the best ones to ask about which rigs hold up in this kind of use (or abuse).

2-17. WHIZBANG DSP RADIOS

The overall function of a receiver is to select, amplify, and demodulate RF signals to an intelligible output. Whether this is accomplished in the analog realm or digitally is mostly transparent to the user. The new DSP radios are touting as having such features as 32-bit floating point DSP Digital IF filter w/ 50 selectable bandwidths, digital twin pass band tuning (PBT), AGC Loop Management, Twin Peak Audio Filters, soft and sharp DSP filter skirts, up to 50+ filter combinations and much more. DSP IF filtering systems can eliminate the need for a several expensive crystal filters

The higher bit rate in digitally processed signals can improve performance over the older 16-bit technology. A big advantage of the DSP radios is noise reduction which can make the difference between hearing the weak DX or not. Be aware that some top notch DXers still prefer the analog transceivers, not quite convinced that the DSP radios have arrived. The quest for a perfect or near perfect DSP design is perhaps still under way. But current DSP designs are constantly improving and there may be one just for your needs.

Some still use the Drake R4C's and T4XC's for low band contesting or the Kenwood TS-830 for an all around DXing/contesting rig. These will require more transmitter adjustments than the newer solid-state rigs but this becomes second nature with use. For those on a low budget these can be excellent starter rigs for under $500. Compare the two versions carefully in light of your budget, anticipated activities, contesting, DXing, and rag chewing. The folks at the DX club can offer lots of advice on the various models.

The new high end HF to UHF all mode radios sell for far less than buying three radios to cover the same bands.

But it remains to be seen if these have compromised the HF capability in order to achieve the "all in one box" approach. You might want to compare these to the DXer and contesters preferred radios as a matter of tradeoff of cost to performance.

2-18. TRANSCEIVER PROGRAMMABILITY

The newer microprocessor radios can be computer-controlled. Some offer controlling two radios simultaneously as well as controlling most radio functions. For those who favor the keyboard over manual control, these can be time saving and offer advantages during contesting and DXing. Fast typists can do wonders with this arrangement. Older radios require additional hardware such as the W1GEE products at URL: http://www.sarrio.com/sarrio/w1gee.html

2-19. TRANSCEIVER OPERATING CONTROLS

Following are discussions of a modern transceiver's operating controls in the way they are used in DXing. These are oriented around Kenwood's nomenclature, for others; check your op manual for corresponding control names. For example AIP is IPO on a Yaseu. RIT is a CLARIFIER on a Yaseu.

Band Stackable Registers. The newer solid state rigs use storage registers to allow ham band selection at the push of a button or by up or down buttons. The up button will take you to the next higher ham band or lower band with the down button. In some rigs, the register remembers the last tuned frequency on that band and certain settings – see your operating manual. On some rigs the ham band switching can be changed to 1 MHz steps – handy for SWL listening.

VOX (Voice Operated Transmit). For voice activation, a circuit senses your microphone audio and when there is sufficient audio (adjustable with the VOX Gain pot); it activates the transmitter keying line so you don't have to use a mike button or a footswitch.

When you quit talking the keying line opens and the receiver audio is enabled. If you leave long pauses between words, the VOX drops in and out, which can be annoying. A VOX Delay circuit determines the interval from the end of a word to the dropout.
Another control for VOX operation is the Anti-VOX control. VOX operations are difficult with other room noise such as a speaker; therefore anti-vox reduces the tendency of the VOX to activate from other sources than your voice.

ATTENUATOR switch(es) are used to reduce the overall receiver gain. When a neighboring Ham comes on the air with a big signal, it can "blow out" your receiver front end causing IMD and lots of strange sounds. Using attenuation or AIP helps minimize the adverse effect of a close by transmitter or extremely large signal.

THRU/AUTO switch bypasses or includes the built-in antenna tuner. Note that in some rigs the antenna tuner is only effective in transmit mode, not receive. The internal tuner may have compensation for VSWR's up to 1.3:1 but not have the range of impedance matching that an outboard tuner offers. But they will do a good job on slightly off resonance antennas and matching to a linear input that is not 50 Ohms.

The latter shouldn't be necessary with a well designed linear, but be aware that some linear manufacturer's designs do not have an input network for each band, typically having only one input network for two bands e.g., 10M and 12m. So there might be a slight mismatch between the transceiver output and the linear input on both bands, usually the internal tuner will compensate for the mismatch.

There have been user reports of some internal auto-tuners resulting in a power loss. This is easy to verify. Load the rig into a dummy load with an external wattmeter between the rig and the dummy load. Switch the tuner in and then out, and check for a drop in power. Makes a good case for an external low-loss tuner or self-resonant antennas.

CW/REV. In most transceivers, in the CW mode, the internal BFO uses the Upper Side Band, in CW/REV it uses Lower Side Band. This is useful during split operation pileups where the mob is calling up a few kHz and getting into your USB passband, particularly the strong big gun stations. Switching to CW/REV can eliminate a lot of this interference.

AUTO ZERO BEAT. This CW feature is found on some of the newer rigs and automatically and exactly matches your transmit frequency with the station you are receiving.

Some DX operators have their receivers set to very tight filtering and it is easy to be outside their bandpass, so be sure to zero beat or auto zero beat the desired CW signal. Manual zero beating varies from rig to rig, so check the manufactures operating manual.

TF SET. This control is discussed in depth in Chapter 5, Split Operations. Basically it allows you to rapidly set or check the transmit frequency during split or RIT/XIT operations without transmitting. Practicing and gaining proficiently with this control will allow you to more effectively work split operations thus yielding more contacts. The TF Set button can also be used to check the repeater offset frequency when operating HF repeaters.

VFO Controls. Modern transceivers have two VFO's and a variety of controls are used to control how they are connected in the circuits. For ease of ID, we will refer to these as VFO A and VFO B. For receiving,
VFO A can be set to a particular frequency and band, and VFO B to the same or a different band and any frequency within that band. This might be useful for tracking two different DX operations, but with the advent of the Quick Memo feature it is of limited advantage. See MEMORY AND QUICK MEMO control below.

The big advantage of two VFO's is the ability to listen on one frequency and transmit on another. During split operations, RIT/XIT (see below) will do this but it has a limited range. VFO controls include equalizing both VFO's to the same frequency. Equalization includes frequency modes, and filter selections. One can select either VFO A or VFO B for receive or transmit. For more on this – see split operations.

AIP. Advanced Intercept Point. A feature on Kenwood transceivers (IPO on Yaesu).This has to do with the 3rd order IMD intercept points. From a user's standpoint it helps reduce modulation products when strong signals are present, see IMD above. Some rigs automatically default to AIP or equivalent for 40 meters and below.

<u>Keypad 0-9</u>. Useful for entering frequencies for storage into memory. Otherwise not often used.

<u>Power Level Control</u> – On solid-state rigs, adjusts the power output from the maximum rated power output down to minimum – 5 watts on my Kenwood. So I am a QRP station at minimum power. When driving a linear – initially REDUCE power so as not to overdrive the linear even though you use ALC around the exciter-linear loop. Tune the linear, then increase the exciter power for the desired linear output, <u>retuning</u> the linear as necessary. See linears below. ALC is Automatic Level Control used for feedback from a linear amplifier back to the exciter to prevent overdriving. Note that some DXers DO NOT use ALC in the exciter-tube linear loop as it may change somewhat band to band. However ALC is mandatory for solid state linears.

<u>Scan Mode.</u> Not very useful for scanning the HF bands. But may help when scanning the 10M repeaters or the 6M calling frequencies.

<u>RIT.</u> Receiver Incremental Tuning (Clarifier on some rigs) allows you to receive slightly above or below your transmit frequency. For DXing, it is typically used for CW split pileups - set your main dial to 2 to 5 kHz up from the DX station and use RIT to hear the DX station.

If an SSB DX station is operating with a small split, the RIT range may allow you to use RIT to achieve the split. RIT typically covers +/- 5 kHz or +/- 10 kHz and is frequently used for CW split operating, but not often in an SSB split operation.

Another use for RIT is when you wish to transmit on a given frequency and the station you are receiving is off frequency or drifts away or starts chasing you by retuning as you tune on them. Rather than change the main tuning dial (thus your transmit frequency) use RIT. The Main Tuning dial determines the transmit frequency and the RIT determines the receive frequency.

<u>XIT</u> (Transmitter Incremental Tuning). In XIT mode, the Main Tuning dial determines the receive frequency and the XIT determines the transmit frequency. It is used when the DX station requests that you call slightly up in frequency (split). Thus you can listen on his frequency (Main tuning Dial) but transmit up by his specified offset (XIT).
This is particularly useful for CW, as one doesn't have to fiddle with VFO settings. On phone however, the split frequencies are typically out of the range of the XIT range so dual VFO's are used. XIT typically covers +/- 5 kHz to +/- 10 kHz and is frequently used for CW split operating. Also try transmitting a little above the DX frequency (on USB) –emphasizes the high tones. Transmit a little below the DX frequency for LSB.

An important feature for contesters is the ability to clear the RIT and XIT with the push of a CLEAR button - saves time for those in a hurry.

<u>NOTCH</u> control (ANF in some rigs). Useful when you are listening to a DX station and a lid throws a sustained carrier on the DX frequency. The notch function will all but eliminate the interfering audio as a result of the carrier. If you are on 40 meters in a rag chew, the band changes and an AM broadcast station comes up, again use the notch control.

IF SHIFT. Shifts the center frequency of the IF filter bandpass but not the bandwidth. Can be very effective in minimizing interfering signals. Shifting the IF center frequency does not change the current receive frequency

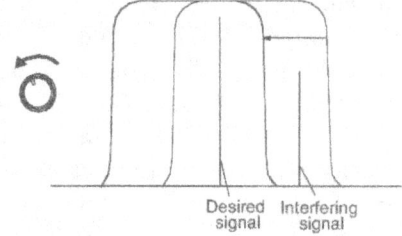

In the drawing above, the IF passband has been shifted to the left so as to center on the desired signal and minimize the interfering signal. Another use of the IF Shift control is to use it as a quasi-tone control. It can be adjusted to cut high or low tones and might give a better audio response to copy that weak rumbly or noisy DX station.

PBT. Pass Band Tuning. Similar to IF Shift but passband tuning, as implemented on the some ICOM models, changes one edge of the passband. Some rigs have twin passband tuning. Check your operating manual for the range of narrowing or widening the pass band.

(VBT Variable Bandwith Tuning) increases or decreases the IF pass-band width

IF FILTERS (Filter Select). The older rigs offer optional filters such as crystal or mechanical filtering. These have sharp skirts and are very effective in minimizing interference. The newer DSP radios may not offer optional filters as the IF bandpass characteristics can be altered digitally. For a detailed discussion of mechanical vs. crystal filters – see URL: http://www.wa3key.com/filters.html

AGC Control. Automatic Gain Control. During a ragchew, engage the AGC control to prevent crashes and noises from coming in between words of the incoming station. Slow AGC is useful for maintaining constant audio output in roundtable conversations with stations of different signal strengths. For DXing and weak signals, use fast AGC and reduce the RF Gain. Using fast AGC prevents the DX cops or sharp static noise from blocking your receiver for the length of the AGC delay. Recommended practice is to adjust RF Gain control until the AGC is inoperative and set the audio as required. When a strong signal appears, as is quite common in a contest, the AGC engages and saves the eardrums and gives time to readjust the controls, if necessary.

The ideal is an AGC threshold that remains constant, so that it overrides the RF gain control at any setting of the RF gain control. This, plus the audio gain setting, minimizes the amount of knob twisting.

When trying to hear weak signals, especially in the presence of strong signals, AGC can be counter productive as the strong signal captures the AGC, reducing receiver gain, which reduces the audio level of the weak signal you are trying to hear.

This may explain the never-ending transceive pileups, especially on SSB, where the DX station is often not heard returning because many pileupees have their receivers set to full RF gain and often slow AGC which maintains a low RF/IF gain as long as the strong signals are calling. Fast AGC with the RF gain control reduced allows you to hear between the syllables of the calling stations, often allowing you to copy the returning DX station.

Some operators operate successfully with no AGC in the CW mode using RF Gain and AF Gain riding. This may work for you in SSB for very weak signals with no big signals around. Also another operating technique is when experiencing heavy QRN is to try no AGC and widen the filters so they respond faster with less ringing.

RF GAIN. Many Hams keep this at max as a matter of course or to read the S-Meter. (The S-meter is only calibrated when the RF gain control is fully clockwise). Many DXers operate with the AF gain turned up and adjust the RF gain until the DX is good copy using a headset. This technique seems to subjectively improve the signal-to-noise ratio and is a lot easier copy than with the receiver gain wide open. Old time CW ops operated this way - works on phone as well. Here we get into the capabilities of the human ear, masking, threshold, etc. But suffice it to state that using your RF Gain and AF Gain for optimum perception works for a lot of DXers.

By using reduced RF Gain and optimum AF gain do we get an improvement in signal to noise ratio?? The theory boys say NO, but the human sensory processes gurus have this to offer: there is more signal and less noise by using maximum AF gain and minimum RF gain. The reason is that lowering RF gain minimizes the amount of atmospheric static, interfering stations, intermod, and so forth for the auditory system to deal with. Hence, the auditory system is subjected to more of a pure stimulus rather than dealing with a masking threshold. Well no matter who is right, DXers use the technique very effectively.

Another factor in the use of the RF Gain control is what might be termed "maximum useable sensitivity". That is to say, with 30 microvolts of noise on the lower bands, do you need a one-microvolt sensitivity??
Nope you can't use the enormous gain of the modern transceiver under these conditions and turning down the RF gain is a matter of course for DXing. In fact some manufacturers insert attenuation into the receiver when switching to the lowbands.

RF PRESELECTOR. In the older rigs, the first RF stage had a fairly narrow bandpass and was tunable to peak up for maximum signal strength. The all solid-state rigs opted for a wideband RF stage where the filter for a particular band covers the whole band and is not tunable from the front panel. Now coming full circle, some of new rigs have a "preselector" which is a narrow band RF amplifier that can be tweaked from the front panel. With double and triple conversion receivers, the wide band approach doesn't seem to have been a hindrance in receiver design.

PREAMPLIFIER. May be termed differently on some brands – such as Yaseu calls it IPO (Intercept Point Optimization). Kenwood calls it PREAMP. In both, the preamp can be selected for higher gain or turned off. When the preamp is on, the receiver dynamic range may be degraded, so when excessive interference occurs, turn off the preamp.

SPEECH PROCESSORS. Processing can be very effective for packing your audio in compressed form. It tends to raise the level of the low amplitude portions of your voice and compress the high amplitude levels. When working DX, this can be very effective, but annoying when rag chewing. Perhaps it should be called DX audio boost. The Processor level control varies the level of compression which can be heard through the use of the MONI function. Also, a nearby DX buddy can assist in setting up the level of compression; non-DXers are a poor bet for helping you to determine your optimum settings. Compression settings over 15 dB can easily result in distortion, so be sure to monitor this setting.

FULL OR SEMI CW Break-In. In FULL (QSK) the transceiver is returned to the receive mode as soon as the key contacts open. Thus it is possible to hear if another station wants to break-in or hear any interference. In SEMI, return to receive is similar to FULL except a delay is introduced as determined by the DELAY control. Note however when using linears, modification may be required for fast response; see linears paragraph below. Transceivers vary considerably in their full break-in characteristics. Ten-Tec has a well-deserved reputation for superb QSK operation. While DXing, with QSK, you can follow a DX-station much more closely and you can stop transmitting the instant you hear the DX-station start to transmit.

QSK operation (full break-in) in CW allows you to hear in between Morse dits and dahs, which is great for Net operation, contesting, and allows the other station to instantly break-in to alert you to slow down or wait for a bit. QSK operation is usually accomplished with silent electronic switching and is essential to high-rate CW operating, since you waste less time. A tip off is if somebody is trying to send at the same time you are. If your contest exchange isn't being received for some reason, you can immediately respond, rather than waiting until you have sent the entire exchange. High-speed traffic handling uses QSK for the same reasons.
You can probably do without QSK for general DXing; it's certainly nice to have, though. But it can be difficult to be competitive as a CW contester without it.

PITCH (CW Function) The default pitch for many rigs is set to 800 Hz; modern rigs have a control or a menu to change this to suit the operator. Experienced operators prefer a lower pitch – experiment for your best listening.

NOISE BLANKERS. These are effective for ignition noise and wide racket noises such as the infamous "Russian Woodpecker (an over the horizon radar)" which hasn't been heard for a number of years. The narrow noise blanker is most effective on short duration pulse noise. The noise blanker gain control should be adjusted for best effectiveness consistent with minimum distortion. Older rigs had a nasty habit of allowing nearby strong signals to blow by the noise blanker, so adjusting the noise blanker gain was necessary.

ALC METER. ALC - Automatic Level Control. A feedback voltage in the transmitter's output amplifier used to prevent amplifier overload. On SSB, the mic gain is adjusted for an ALC meter reading in the ALC meter safe zone. In CW, the Carrier Level is adjusted for the same reading. Manufacturers tend to have this set on the conservative side, so a better check is with a monitor scope to see where you really want to set it, consistent with no distortion. For side band you need to pack in the audio for maximum punch as one can hear lots of under modulated signals on the band. Also have a DX buddy check out your settings.

CARRIER LEVEL. Usually functional on the CW, FSK, and AM modes but may be used as part of the speech processor circuits. This control affects the power output of your transmitter and should be adjusted as per the manual which is generally set so that the ALC meter reads within the limits of the ALC zone. Note that in some rigs this may vary from band to band – so it should be checked as you operate on different bands.

MONITOR function is handy for checking what your audio sounds like with headphones. It's a good way to check the speech processor settings. Also it can allow you to hear if you have RF feedback – you'll know it when you hear it (rattle rattle). Besides using Monitor to hear your signal, have a local DXer check your Mic and Processor settings. Don't rely on the average Ham to assess this, many will object to the processor no matter how it is set. An experienced DXer can assess your settings, as they are experienced in hearing processor signals in a DX situation.

MEMORY AND QUICK MEMO. Memory channels are convenient for storing oft-used frequencies such as nets, RACES, etc. The quick memo is really neat for DXers. Suppose you are listening to a rare DX station and he/she is working sequentially by districts, you're a W4 and they just finished 4's and now on 5's – ya got a long wait before he/she gets around to you again. So you use the quick memo to store them away, including frequency, filter and VFO information (on the newer rigs) and tune around for other DX. Periodically you can hit Quick Memo and check for what call area is now being worked.

You're an old time DXer and nothing is on except that new DX entity, you can use Quick Memo to contain all of the announced frequencies and bands that they will operate. By using Quick Memory Recall you can check all the bands quickly.

QUICK TUNE. Some transceivers offer a control to tune in steps of 5 or 10 kHz instead of spinning the MAIN Tuning dial. This is handy for quickly tuning up and down the band. On Kenwood radios this is the MULTI/CH control. In the quick memo mode, this control also allows you to select the quick memo channels.

MAIN TUNING DIAL. Many rigs allow adjustment of the drag on the main tuning dial so you can easily spin the dial to get across a band or choke it down a bit if you are heavy handed. An easier way to get up and down the band is use the Quick Tune control set for either 5 kHz or 10 kHz increments. Since SSB signals are 2 to 3 kHz wide, using a 5 kHz increment will allow you to step up or down the band and hear most signals present. Also using the FINE Tuning function will allow for easier precise tuning on a signal. Some Main Tuning dials have a thumb detent for better control. On later model rigs, when turning the Main Tuning, the tuning rate increases or decreases with the speed of the spin that is applied, making it easier to rapidly approach the desired frequency

As one tunes across the bands and listens to QSO's or pileups, it is amazing how many times one of the stations is improperly tuned for an SSB signal, being way too high or low and off center considerably. Practice tuning in on different voices, bassy, high-pitched etc. Some DXers call slightly off frequency purposefully to give a higher-pitched sound.

MENUS. The microprocessor radios have numerous menu items that can be intimidating at first, but most of these have default values that are the most used settings.

However, the new radios have some really neat menu features such as boosting or cutting transmit audio, setting AGC ranges, CW weighting, CW pitch tones, display brightness, frequency tuning steps, scan modes, DSP settings and much more. One should not be intimidated by all the menu items and bells and whistles on the new radios. After they are all set up, you might use only 6 to 10 controls.

It is wise to read the entire operating manual, because the new radios are not your father's Hallicrafters – for sure. Once you have set up all your menu items, you might use a dozen or so front panel controls at most and not have to bother with the menus very often. Some radios have two main menus A and B. This is neat for setting up multiple operating modes. For example, one can be set up for rag-chewing and the other for DXing. For rag-chewing, one can select a full audio range, where for DXing, high treble boost might be used. AGC delays for rag-chewing might be slow, but fast for CW or DXing.

There are lots of world wide 10-meter FM repeaters and these are fun to work on skip. The menus in the new radios allow for repeater offset and PL tones as well. The local 10-meter and 6-meter repeaters are a great place to have semi-private QSO's without the constant interruption that may occur on a 2-meter repeater. Also DXpeditions have been showing up on 10M FM.

SECRET MENUS. With the advent of the microprocessor radios and its attending menu selections, some transceivers have "Secret Menus". Some of these are for factory servicing, but others are for operation setup. The Internet is a good place to check for these. Try google.com and type in your model number. Beware of fiddling with items you don't understand, as some can be very obscure. My late model Kenwood radio has a variety of these goodies, some useful, most not.

DUAL RECEIVE. Some of the more expensive rigs have Dual Receive capability; this is especially useful for contesting. Dual receive for contesting allows you to tune the second receiver in between CQ's and pop a new contact when you find another guy calling CQ.

Also, for rigs that can receive another band (TS950 receive same band only, FT-1000D receives same or another band) you can either pop someone calling CQ on another band if you don't violate a 10 minute stay on one band or 6 multiplier per hour rule). Also, you could check propagation on one band while you're working another. Lastly, you could use the second receiver in a DX pileup when the DX is operating split and follow where the last person worked was at - instead of having to shift A/B VFOs on a single receive rig and not have to keep punching the T-F button on Kenwood rigs like the 940 and others.

2-20. YOUR SIDEBAND AUDIO

For DXing it is vital to have full modulation, optimized speech processing, and equalized audio.

While tuning through the bands, you can hear a lot of weak and puny audio signals, because folks are afraid of over-modulating, so they under-modulate. Here are some procedures that have worked well for many DXers. Have a savvy local station check your signal during these setup procedures. The station checking your audio should be familiar with good full modulation and good sounding speech processing, many HFers are not, get a savvy DX buddy to check you out.

With RF POWER set to max, SSB mode, and with the processor OFF, adjust the MIC level as per the Instruction manual so that on voice peaks, the ALC is proper.
Some rigs are very conservative on the ALC meter; so have a DX buddy check for maximum modulation, but not splattering, distorted, or excessive bandwidth.

While testing, use test "1, 2, 3, 4" and have your buddy assess your audio. Back off the mic gain a bit from the maximum clean modulation level. The optimum setting may not agree with the ALC meter, but you want full modulation, not under-modulation, so on-the-air tests or a bandscope will help achieve this.

With the processor ON, rotate the PROCESSING control fully clockwise. Speak normally into the mic with a test OHHHLLLAAAAA and adjust the RF POWER level for maximum output. If you use headphones and the MONITOR function, this will be pretty grim sounding audio, but wait – now turn down the RF POWER control until the RF power output drops by 10%. With most transmitters, reducing the peak power by about 10% from the maximum (saturated) level makes a significant improvement in IMD. From here on out, leave the RF Power control set as above.

Listen to your signal on another receiver and/or have a DX buddy check your audio. Adjust the PROCESSING level control fully counter-clockwise, and then adjust it until your voice sounds loud and crisp, but not distorted. Compensate for shack background noise. Have a savvy local station check your signal for splatter and bandwidth and reduce the PROCESSOR level accordingly.

For equalized audio, if you listen to pileups, the signals that seem to stand out are the YL's, the kids, and audio that has a fair degree of crisp high notes.

Tailoring your audio can be a real advantage in the pileup rumble. May not beat out the big guns, but it sure helps in the long run. Reduced lows and some mids with boosted high frequency tones are the key to a punchy signal.

The Astatic D-104's are famous for their high-frequency characteristics, rising to about 10 dB at 3,000 Hz. For today's solid-state radios, an amplified D-104 is recommended as it provides a proper impedance match. Bob Heil has perfected the art of audio for Amateur Radio and offers a line of microphones for DXing or rag chewing or both (HC-4 and HC-5 elements). Reduced lows and boosted highs are the characteristics of the famous DX cartridge.

The new digital radios (Kenwood TS-870) offer built-in tailoring for transmit audio. This feature is very powerful for adjusting your audio to suit your voice and working conditions. Features include provision to adjust the Processor low and high frequency response, Transmit Bandwidth and Transmit Equalizer. Although the effects of audio can be heard in a headset using the Monitor function, a much better method is to monitor your audio with a nearby receiver and a good quality headset, then you'll be sure of what your audio sounds like on the air. Or have a savvy DX buddy assess your audio for punch and no distortion.

2-21. MODIFICATIONS

There are a dozen or so sites on the Internet that specialize in radio modifications, everything from installing new filters to extending frequency ranges (MARS or CAP). Be aware of the FCC rules regarding modifications and the aspect of voiding manufacture's warranties. Some of the mods are for improving or correcting problems of earlier serial number radios. Modern radios contain very small parts and surface-mount circuitry and are difficult to work on, so a very light soldering iron, magnifying glasses, and steady nerves are a must. Radio mods on the Internet can be found at URL: http://ac6v.com/techref.htm#MODS

2-22. LINEAR AMPLIFIERS

Although it seems that the name of the game is a big linear, most will tell you to put your money in a good antenna first. A good antenna will not only let you be heard over the pileup, you can hear signals that you might not discern with a marginal antenna. Although an amplifier may boost your signal output, it won't help you receive any better. Once that is in place or the HOA limits you to a vertical in the trees, a linear will definitely help, but maybe not as much as the figures might imply.

For example, let's start with a 100-Watt exciter then add an afterburner of 800 Watts. Doubling power is 3 dB and an S-unit is 6dB. So from 100W to 200W is 3dB, 200W to 400W is 3 more dB, and 400W to 800W is 3 more dB for a total of 9 dB or only 1-1/2 S-Units possible increase on the receiving end. One can achieve almost the same result with a 3-element beam. In terms of cost however, it may well be more cost effective to buy a moderate size amplifier (800 to 1200W) which provides 9 to 11 dB gain on ALL BANDS than to go for an equivalent improvement in gain with bigger antennas on ALL BANDS.

Although one and a half S-Units does give a little more oomph while working a pileup, it really comes into play for weak signal work where if it gets you just up out of the noise on the receiving end, two way QSO's may be possible where without the linear, the DX just can't copy.

Good signals can be achieved on 10 meters with 10 Watts of power if the propagation is good to very good. But it is a different story on the low bands. Here atmospherics are severe and higher power transmitted signals are in order.

For six meter work, using an amplifier is especially attractive for working the weak signal propagations, such as aurora, scatter modes, and multihop sporadic E.

And 20 meters is kilowatt alley – tough for the low power stations to compete but possible. When propagation is good on 20 meters, QRP levels will get through if the band QRM is low. I have worked Europe with 6 Watts and a vertical on 20 meters, so it can be done, but much easier on 10 meters though when this band is open. Remember that propagation conditions play a major role in getting signals across the globe. Under the right conditions, 6 Watts will get half way around the world.

On the lower bands (160, 80 meters) an amplifier is a virtual necessity. With the noise levels on these bands and the nature of propagation, running 100 Watts is a difficult proposition.

On the upper bands, you can get along with running barefoot, depending on your antenna, indeed many DXers operate QRP and run up enviable DXCC credits. On 10 meters, with high solar flux and the propagation in your favor, you can beat out the big guns back (east, west) running a 100-Watts while they are using a "full gallon".

Connecting the transceiver to the linear can be tricky; some transceivers may have wimpy relays and may require an outboard relay to handle the currents. Be sure to read both manuals for the interface requirements. Be especially careful when soldering up a plug for accessory operation, any shorts on the pins can result in grievous damage to the rig. Use an Ohmmeter to verify integrity of your connections. If in doubt consult with other DXers about your equipment interface.

Running a full legal limit amplifier is best accomplished with a 220 Volt AC line. Running off of 117 Volt line can be dicey, straining the power wiring, and possibly resulting in some distortion to say nothing of house light dimming.

Follow the manufacturers directions for dipping and loading an amplifier, then to achieve a cooler running amplifier with a cleaner output, once you have tuned for maximum output by dip and load, reduce the grid current, by increasing the loading. Loading is increased by reducing the load capacitance, this properly couples the RF to the antenna. This will reduce the grid current considerably with a small drop in RF output power - about 10 percent. The drop in RF output is a small price to pay for lower grid current, which will greatly increase the life of the tube(s), especially metal/ceramic type tubes. The drop in grid current should be somewhere in the area of 30-50%, (i.e., full drive grid current of 200ma, can be reduced to 150-100ma.). If you turn the load control one direction and the RF output peaks and the grid current remains high, turn the load control in the other direction until the RF output drops slowly and the grid current drops quickly.

Some use the duty cycle method of tuning an amplifier; using a commercial pulser or cricket (Centaur Electronics, NI4L, and MFJ makes them), or a CW keyer set for 40-50wpm, this reduces component stress and overheating and excessive power supply loading.

You can reduce power by simply reducing the power output from the exciter with only a small loss in overall efficiency, although "text-book" way is to re-tune after reducing the power.
If you re-tune for the lower power and later decide to run full power, you will need to retune for the higher power level

QSK operation (full break-in) in CW allows you to hear in between Morse dits and dahs, which is great for Net operation, contesting, and allows the other station to instantly break-in to alert you to slow down or wait for a bit. Some use a fast acting reed relay at the linear input and a vacuum relay at the linear output. QSK operation is usually accomplished with silent electronic switching and is essential to high-rate CW operating, since you waste less time. A tip off is if somebody is trying to send at the same time you are. If your contest exchange isn't being received for some reason, you can immediately respond, rather than waiting until you have sent the entire exchange. High-speed traffic handling uses QSK for the same reasons. You can probably do without QSK for general DXing; it's certainly nice to have, though. But it can be difficult to be competitive as a CW contester without it.

AG6K is a guru on this topic and has a web page at URL: http://www.somis.org/ He has several articles in QST magazines as well. Talk to the contest CW operators for specific models of equipment.

Also be aware that some linears do not have a tuned input circuit for each band, but rather have only one for each two bands e.g., 10 M and 12 M. This can result in a mismatch between the transceiver and the linear and increased VSWR. Most internal tuners will compensate for this however.

2-23. SELECTING POWER AMPLIFIERS

The FCC's power rules are based on peak RF output and you need to evaluate the amplifier in that light to be sure your investment is worthwhile. Check the manufacturers' specifications and ask what their power specs are in terms of peak RF power. The duty cycle specification is an amplifier's ability to tolerate continuous hard use. If you are not a contester or don't use continuous carrier modes like RTTY, you may want to lighten up on the duty cycle specs. But an amplifier that yields 1.5 kW PEP or carrier output with no time limit will cover all bases.

The transformer design and weight provide a clue on power rating claims. The top manufacturers use core grain-oriented silicon steel such as Hipersil® or Unisil®. A core of tape-wound Hipersil® or Unisil® with triple-insulated windings yields superb reliability.

Another important consideration is the frequency coverage of an amplifier. Since the FCC prohibits U.S.A. amplifiers to be operational on the 24 - 29.7 MHz range (as shipped), you need to determine how easy or complex it will be to activate these bands. You also need to determine what the performance will be after activation, power output, tuned input circuits, etc.

If you intend to work the WARC bands (and you should, lots of DX there), beware of older amplifiers that won't cover 17M and 12M. Modification could be sticky or impractical.

Many of these can be "horsed" down to 17M, but 12M presents special problems as components may resonant undesirably and blow the amp. Best check with owners of the model you are considering, as mods can be difficult.

Check whether the amplifier has tuned input circuits for each band. Some manufacturers skimp on this by having only one tuned circuit for two bands, e.g., 12 meter/10 meter. Some of these are tunable as a compromise but it is just that – a compromise and you may have some VSWR between the exciter output and the amplifier input.

Other considerations are; step-start inrush protection, separate transformer for the screen grid supply, screen grid current regulation, harmonic suppression, full time metering capability, including plate current, plate voltage, screen grid current, forward and reverse power, and continuous peak reading power output monitoring. Protective circuits include control grid current limiting and plate current over-trip circuits. The use of cooling fans vary from chimney fans to just blowing air across the tubes, and inadequate cooling can be a nightmare in a contest or sustained RTTY operation.

For sustained operation such as a contest or RTTY, research the duty cycles of the linear. The linears with sweep tubes are highly stressed in this kind of environment and may die during the contest or probably at a young age. Some of the commercial amplifiers have inadequate cooling for contest use - however this problem can usually be overcome by fitting an extra fan or new fans. Some amps have had problems with power supplies and other issues – so check with owners.

2-24. SOLID STATE AMPLIFIERS- TUBES VS. FETS

With the introduction of RF Power FETs, one would think that they would eventually replace bipolar transistors and vacuum tubes in HF power amplifiers. Since RF power FETs operate at upwards of 50 Volts, FETs have not replaced bipolar transistors in 12V mobile applications. Also a pair of FETs that can produce 1200 Watts PEP at 30 MHz will cost about six times more than electron tubes for the same application. With FETs requiring 50V at 50A, i.e., 2500W, there is a lot of heat to be dissipated.

Designing adequate cooling is demanding as compared to vacuum tubes since tubes can survive high surface temperatures that would totally frazzle silicon devices. Also, power supply requirements are much more stringent for solid-state devices. Solid-state rigs are unforgiving of SWR and have circuits to reduce power should high SWR occur. They are typically instant band-switching, no-tuning, instant warm-up devices which is a godsend for quick band changing. See URL: http://ac6v.com/radiodealers.htm   For amplifier manufacturers and lots of info.

2-25. HF ANTENNAS

First, to explode an antenna myth, antennas do not have to be self-resonant or have a 50-ohm impedance feed point! The use of long wires, tuners, ladder line, tuning stubs, and other methods have been used for years and will allow for efficient operation.

Let's take a moment to look at antennas for the beginning DXer. Most new Hams put up their first HF antenna (HOA allowing) by walking outside and taking a gander upwards and figure "lets put a dipole on the 20-foot roof". Or maybe string one between some 18-foot trees. Well this may get you across the state but it's not too good for East Africa. So what is the difference? There are a lot of factors but angle of radiation plays a major role as do gain of the antenna, front to back ratio, etc. Angle of radiation is important for achieving long skip distances.

For a lightly ionized atmosphere, the high angle radiated signal may go right thru the layers with little returning to earth. But the same signal squirted out at a low angle of radiation may have sufficient "bending" to get back to earth.

Gain, of course is squishing energy into a narrower pattern, and it increases your effective radiation power (ERP). You can't get out more than you put in, but a directional antenna will concentrate the power into a narrower pattern, hence "gain" compared to a reference (like a dipole antenna or isotropic reference). In addition, the front to back ratio can be very effective in receiving signals from the desired direction and minimizing others.

In a nutshell, on the higher HF bands, you generally want a low angle of radiation, so a beam at 48 feet or a vertical over good ground or with lots of radials will outdo a dipole or even a beam at 20 feet. On the lower bands, angles of radiation can be much higher, so it's a different ballgame. Also on the lower bands, higher power outputs are in order and special receiving antennas, such as beverages or small loops are employed just to be able to hear through all the noise. For low band DXing, John Devolderes' book "Low Band DXing" is a must.

A whole book could be filled with antenna theory (and has been many times), so it is not the intention of this guide to repeat the information. Rather we will look briefly at the types used by DXers and the appropriate references. User reviews can be found at http://www.eham.net/reviews/

Ready for a surprise? Well, according to an ARRL survey "What antenna do you use most on HF?" came up with the following results (total votes: 6589). This takes into account all Amateurs DXers and non-DXers alike.

- Multiband vertical - - 24.4 % (1607 votes)
- Triband Yagi   -- 19.5 % (1287 votes)
- Other   -- 17.8 % (1175 votes)
- Low-band (3.5 to 10.1 MHz) dipole -- 17.5 % (1154 votes)
- High-band (14 to 29.7 MHz) dipole -- 13.7 % (903 votes)
- Monoband Yagi   -- 4.3 % (284 votes)
- Quad   -- 2.7 % (179 votes)

For DXers, the 425 DX News came up with these results

- Beam/Yagi -- 52% (115 votes)
- Vertical   -- 16% (36 votes)
- Dipole   -- 15% (34 votes)
- Quad   -- 9% (20 votes)
- Loop   -- 4% (8 votes)
- Other   -- 2% (4 votes)
- Sloper   -- 1% (3 votes)

2-26. HF ANTENNA HEIGHT

For the first antenna, the beginning HFer might string a horizontal antenna up about 20 feet. After a lot of work, the testing begins and after a few adjustments for SWR, the reports from upstate are satisfying, then the real world settles in and the budding DXer finds it just doesn't work much DX. Strange, it works more or less OK on 10 meters but 20 and 15 meters are a bust. Ah ha – sighs the QRPer – I need a beam or quad on a big tower. Yep that will do it, we reassure him – but what is the DX antenna game all about.

So what's needed for DXing? The secret to DX antennas is pretty much a proper angle of radiation for the band of operation and desired coverage. Higher may or not be better as the angle of radiation is affected by height, band, terrain and other factors.

A local DXer has three bays of antennas (20, 15, and 10 m each bay) at 3 different heights and he can select one for best reception or select two or three with phasing. Most Hams get along with a single tribander (or 5-bander) at a height of 42 to 100 feet. See Chapter 3 for antenna modeling programs.

Low-height horizontally polarized antennas are affected by ground reflection, and typically have higher angles of radiation with respect to the horizon. Maximizing low angle radiation may be desirable for distant line of sight communication links, but not for sky wave (skip) communication links.

For sky wave communication links the most desirable angle of radiation will depend on the layer of the ionosphere being utilized for reflection, and the overall distance of the communication link (radio wave arrival/departure angle).

In some instances a relatively high angle of radiation may in fact be more desirable than a low angle of radiation. Stateside HF net operation, and six meter E-skip propagation are two examples where low angle radiation is usually undesirable.

Some modeled figures for a dipole at various heights above ground

0.1 wavelength high -- Launch angle 90 deg	1.0 wavelength high -- Launch angle 14 deg
0.3 wavelength high -- Launch angle 50 deg	2.0 wavelength high -- Launch angle 7 deg
0.5 wavelength high -- Launch angle 28 deg	4.0 wavelength high -- Launch angle 4 deg

For DXing, typical values for the angle of radiation for the HF bands using horizontal antennas are; 5 to 14 degrees for 10M (antenna height above 34 feet); 7 to 20 degrees for 15M (antenna height above 38 feet); 10 to 25 degrees for 20M (antenna height above 40 feet); and 12 to 40 degrees for 40M (antenna height above 45 feet). These are minimums and in general the angle of radiation decreases as the height of the antenna increases. So all those big towers are to achieve the lower or proper angles of radiation required for DXing. Don't be fooled by the low height antenna that bombs into an up state location; getting to deep central Asia is a different story.

One DX guru advises no one single antenna will cover all the angles supported by the ionosphere due to pattern nulls for horizontally polarized antennas. It takes antennas at different heights to effectively cover all the angles. For low angle radiation of 10 degrees or less you need to be at least 1½ wavelengths high. Is it worth increasing an existing support tower ?

According to a guru, a general rule on height increases is that 50% increase is worth the increase. Less, don't worry about it doing it. On the other hand, low angles of radiation can be achieved with a ground-mounted vertical. However, the vertical has low gain, is omni-directional and prone to man-made noise.

On the technical side, the takeoff angle (TOA) in degrees = $\sin^{-1}(a/4*H)$. Where A = 1st lobe, A=3, 2nd lobe...nulls are the even numbers 2,4...go as high as desired. H = antenna height in Wavelengths. Thus TOA= inverse sin of the quantity $(1/(4*Ant\ Height\ in\ Wavelengths))$

If the antenna has multiple lobes, use A=1,3,5,7,9 (this will only occur above 3 if the antenna is very high). Note that TOA is not a line, but a cone of radiation.

It's also easy to get lost in angles and forget amplitude. You can have a low TOA, but a poor signal because the antenna could be quite inefficient.

For comparison, you always look at the actual amplitude of the signal at a given angle. A half wave up dipole will typically put out more amplitude at a low angle than a vertical over typical ground, but the vertical is supposedly the "low angle radiator". See W7EL's program for Eznec.

Antenna Placements. Often asked is what effect is exhibited when antennas are close to one another ?? One expert advises, measure the antenna impedance with no other antennas around it. Then measure its impedance with other antennas in the vicinity. Any change in impedance indicates mutual coupling. For multiband antennas – this means impedance measurements on each band.

Here's what EZNEC states about the 5 element 20m Yagi that is included in the EZNEC package.

Height	Gain	Take-off-angle	Vertical Beamwidth
30'	13 dBi	25 deg	30 deg
50'	14.5 dBi	18 deg	20 deg
70'	15 dBi	13 deg	14 deg

2-27. YAGI BEAMS

The Yagi antenna was the original invention of Hidetsugu Yagi and Shintaro Uda. In 1926, Yagi and Uda jointly published a paper describing the new antenna. It combines a single driven-antenna with an array of passive elements, called directors that radiate after receiving power from adjacent elements (forming what is called a closely-coupled parasitic array).

This design makes it possible to achieve high directivity with a fairly compact antenna. Important characteristics of beams are directivity (gain), front to back performance, beam width, wind load etc. These can be found in manufacturers' data sheets. For good low angle radiation the Yagi should be at least a half wavelength in the air. For 20 meters this is 33 feet. However, the optimum angle of radiation for 20 meters is in the order of 10 to 25 degrees which equates to an antenna height of at least 40 feet, preferably more. A good source for Yagi beam characteristics is Bill Orr (W6SAI) "Beam Antenna Handbook". Shows radiation patterns vs. height.

Typical gains for a two-element Yagi beam is 5 dB, three elements yields 8 dB, etc. Gain can be referenced to an isotropic radiator or a dipole, so beware of antenna gain figures that don't specify and compare gains with like references. Other important characteristics are beam width (60 degrees), front to back ratio (25 dB), power handling capability, wind load, square area, and others necessary to make an installation. Although the triband trapped Yagi's are very popular, monoband Yagis are essential if you want to be a big gun. An analogy for yagi gain and front to back ratio is squeeze a balloon and pinch the back so only about 1/50th of the air that used to be there is left. The rest gets distributed towards the front, and makes the front side more than a factor of twice larger. But the ratio of the amount of air in the front to that of the back is very large --say, 100. That is – a 20 dB front to back ratio.

With the advent of computer-designed antennas, the newer trapless designs can rival a monobander. See "Publications" at URL: http://www.championradio.com/

One noted contester recommends that the Yagi be as high as possible e.g. 40m Yagi - min.height 90', 20m Yagi = 70', 15m Yagi = 60', 10m Yagi = 40'. Note that with 10m Yagis, they should not be too high - the main lobe of the 10m Yagi can split up, which is an undesirable effect. The gain of a Yagi depends mainly on the boom length, the number of elements being of less importance as long as the antenna has four elements. For protection during high winds, when you can tell which direction the winds are coming, or a storm is approaching, point the element ends into the direction of the approaching winds. Wind load on the boom is minimal compared to 3 or more elements facing the wind. You can use a modeling program to verify this or to determine your dream configuration. On receive, a trapped Yagi may be similar to a monobander – but on transmit the difference really shows up.

For cleaning aluminum tubing, an antenna guru recommends the use of #0 steel wool, along with dishwashing detergent. First wet the aluminum tubing with water and wet the steel wool. Add a small amount of dishwashing detergent to the steel wool and then scrub the aluminum. Rinse out the steel wool periodically with water. Then add more dishwashing detergent to the steel wool and clean again. Rinse off the aluminum tubing periodically. Other methods are Scotch Bright, vinegar and baking soda, wet and dry emery cloth - 600 grit, Cameo Aluminum Cleaner. After cleaning a piece of aluminum tubing, rinse it thoroughly with clean water and then dry it off with a clean rag. Unless you totally disassemble and clean the traps, never attempt to clean the outer aluminum shells of antenna traps as water and steel wool fragments can enter the traps causing shorts and excessive trap moisture content. Another suggestion is to clear coat the assembled antenna with clear spray paint. This also goes for the antenna connections as well. Tape them well with scotch 88 electrical tape and then clear coat the assembly. The paint seals the aluminum from the elements and seals any tiny holes that may be in the wrapping job.

When assembling any aluminum antenna, put a thin layer of based-based anti-seize compound on all aluminum-to-aluminum joints. These are Penetrox-A from the Burndy Corporation or another is Noalox. Never varnish any antenna - the varnish penetrates into the joints and causes continuity problems. Painting antennas for stealth can be done. Disassemble the antenna, apply paint, then scrape away any paint where the aluminum makes an electrical connection. When restoring an old antenna, consider replacing all the hardware (nuts & bolts) with stainless steel. **Cutting slots in aluminum tubing**, gurus recommend a circular saw with a thin wheel grinder/cutter.

2-28. HF VERTICAL ANTENNAS

The vertical antenna is desirable for those not able to put up a tower. Two major types are the ¼ wave with ground radials or elevated radials and the newer so called ½ wave verticals. Also co-linear antennas or phased array verticals can be used for increased gain

With the ¼ wave unless you have excellent ground conditions e.g., over salt water or a marsh, be prepared to put out a slew of radials, 24 or more for ground mounting.

By mounting the vertical up in the air, you might get away with 4 radials per band, as elevated verticals work well with fewer radials. If you elevate the feedpoint to at least 15 feet, four radials will perform just as well or better than a ground mounted vertical with 120 buried radials. This was the conclusion of K8CFU who took hundreds of field strength measurements from both configurations (Radcom, Technical Topics April 89). Consider one pair of radials routed in opposite directions. The currents in those two radials are out of phase and thus the fields cancel at a distance.

If you elevate those radials high enough, the fields cancel before they encounter the ground and thus avoid a lot of ground loss. With the radials on the ground, the fields can't cancel before encountering the ground and thus there is a considerable amount of ground loss, so you need lots of radials to try to mimic a perfect conducting plane.

One antenna guru has this to say, "Earth loss can reduce the antenna's efficiency by as much as 50 percent. Earth loss is caused by your antenna's ground return, (the shield side of your coax); being connected to soil that has less-than-perfect conductivity. The earth acts as a large resistor, or RF sponge soaking up all your precious power. Is there a way to increase soil conductivity? Yes just add wire radials at the base of your vertical. Arrange them in a spoke fashion, add as many as you can - up to 120 of them -- which is what most AM broadcast stations use.

If you want to get as near to "no loss" as possible, you will have to install at least 100-halfwave radials, (that's 32-feet per radial for 40- meters by the way). Most of the time, that many radials and the length just isn't possible, so just install as many as you can, or consider the ½ wave vertical". For more on radials for ¼ wave verticals – see Butternut write-up at URL: http://www.bencher.com/pdfs/00803ZZV.pdf

One vertical guru has this to recommend for installing radials. Use anywhere from #12- to #16 stranded and double coated house wire to resist breakage and corrosion. Users report having these type of radials in the ground for up to 14 years and still in like new condition. For "on the ground surface" installation, tack them down every 10 feet. Over time, the grass will grow over the radials and the lawn mower blades will be well above the wires. For on the ground installation, schemes are aluminum stake pins available at hardware stores, or with U-shaped wire fasteners made out of old metal coat hangers. For 'buried radials", installations use a lawn edger, roto-tiller, or a chain saw to create the grooves.

Studies have shown that radial effectiveness diminishes as the radials are buried. Experiments on 160 meters have shown that radials in the form of a counterpoise 15 feet above ground produces the best far field strength, followed by a counterpoise 5 feet above ground, then radials on the surface and then radials installed in the ground. Your results may vary depending on ground conductivity.

The "chicken wire" approach could potentially cause some problems over time, if you use inexpensive material, as it will corrode ... also, joints between sections should be bonded well electrically, not just overlapped. Professional installations use copper mesh -- not galvanized chicken wire.

The newer ½ wave commercial verticals do not require radials; in fact it screws up the antenna if you attempt it. They are ½ wave end fed (high impedance point) with a matching box.

The best explanation I have had regarding no radial ½ waves is when you base feed a ¼ wave vertical antenna; you connect one conductor of the feedline to the antenna and the other to the ground system. The same amount of current that flows into the antenna also flows into the ground. The current flowing in the ground results in power loss. The efficiency of the antenna is a ratio of the radiated power to the input power. Assume a resonant base-fed quarter wavelength vertical has a radiation resistance of about 36 Ohms, so the current is relatively large.

A base-fed resonant half wavelength vertical, on the other hand, has a radiation resistance of at least several hundred Ohms. Assume a poor ground system (a few radials) with a loss resistance of 36 Ohms. The efficiency of a quarter wavelength vertical would be in the order of 50%.

A half wavelength vertical with a 1000-Ohms radiation resistance connected to the same ground system would have an efficiency of over 90%.

You do need a ground system for base feeding a half wavelength homebrew vertical, but a ground rod will help. If you center feed it, you don't need a ground system at all. As mentioned before, do not use a ground system with a commercial ½ wave vertical like the Cushcraft series.

The radiation pattern depends on the ground characteristics some distance from the antenna. Over average ground, it's very nearly the same as a quarter wave vertical at low angles (perhaps a dB stronger, neglecting efficiency differences) but considerably weaker at higher angles.
You can easily see the differences with the EZNEC demo program. See URL: http://eznec.com/index.shtml

In a nutshell the ½ wave radial-less verticals do work, in general giving a good low angle of radiation, but like any vertical they are prone to man-made noise and radiate "equally poorly in all directions" but hear everywhere. If you have excessive power line noise, a call to the utility company is in order. They are happy to find leaky components and insulators – and many times is to their benefit as well as yours. Folks like to compare the ½ wave verticals to the ¼ waves claiming the ¼ wave far outperforms the ½ wave, but they neglect to tell you of their 32 radials on the 40M band!

Shortened verticals seem like a good idea until one looks at the efficiency figures that might be encountered. The efficiency of the quarter wave vertical antenna is the ratio of its radiation resistance to its total feed point impedance. Assume an antenna at a full quarter wavelength with ground losses such that the efficiency is 40%. Now build another 40-meter vertical but shorten the physical length by half by adding a loading coil somewhere along its length. The coil lengthens the antenna electrically to resonance but at a sacrifice in efficiency. Without crunching the numbers, this shorty will have an efficiency of about 12% and that's why full 1/4 to 5/8 wave antennas are better if you have the room.

A good publication for vertical antenna comparisons is at "Publications" URL: http://www.championradio.com/ One antenna guru states a dipole should be a minimum of 3/8 Wave Length.

2-29. CUBICAL QUADS

The Missionary Radio Station, HCJB, at Quito, Ecuador, is the birthplace of the Cubical Quad antenna. To effectively broadcast to North America, a huge 4-element parasitic beam was designed, built and erected with great effort and centered upon the heartland of North America. The big beam added a real punch to the 10kW, 25-meter station. However a totally unexpected phenomenon occurred from the environment of Quito.

Operating the high-Q beam antenna in the thin wet evening air of Quito (10,000-foot altitude) caused gigantic corona discharges from the tips of the driven element and directors. The heavy aluminum tubing used for the elements glowed with the heat of the arcs and large molten chunks fell off of the antenna!

Clarence C. Moore, W9LZX, was one of the engineers at HCJB assigned to solve this problem. The main criterion was to design an antenna with no ends that could be subject to discharge. A concept evolved into designing a folded dipole with the loop pulled open.

This loop later became the quad design. A quad antenna with a reflector was fabricated and erected and as the evening dew collected on the antenna wires, no arc over or corona occurred.

A 2-element quad will have a gain about the same as a 3-element beam. And a quad is a full-size antenna, not trapped or shortened boom length. Quad lovers claim the quad hears well before other antennas – a band opener and band closer. But beam lovers dispute this. Quads are allegedly quieter that other types, so the quad lovers say. An excellent source is Bill Orr's antenna book on quads, homebrew info as well. Tells how to maximize gain, or optimize front to back and the sticky problem of matching the feed line to the driven element.
One can maximize gain by adjusting element spacing but front-to-back ratio will suffer, so it becomes a trade off between these two factors. Another legend is that quads will operate at a lower height than a Yagi for the same performance – there is little evidence to support this as demonstrated in antenna modeling.

For multi-band quads, ideally, the boom is in spider fashion so as to give optimum spacing between the driven element and the associated reflector and director.

Using a remote-controlled coaxial switch to switch between driven elements solves a lot of the problems associated with common ties.

Since the quad has a balanced feed-point, the use of baluns is recommended. A balun is a contraction of <u>bal</u>anced to <u>un</u>balanced. The device matches an unbalanced transmission line to a balanced device such as an antenna. CQ magazine has a book by Jerry Sevick,W2FMI, called building and using baluns and ununs...price $19.95.

A quad at ¼ wave length above ground might have a radiation angle of 40 degrees, at 3/8 wave length – 32 degrees, ½ wave – 26 degrees, 5/8 wave – 23 degrees, ¾ wave – 18 degrees. These are also approximate figures for Yagi antennas.

In high wind or areas with icing conditions, a quad may require more maintenance than a strong beam, so talk to locals to gain their experience.

A recent publication on cubical quads is by L.B. Cebik, W4RNL's "Cubical Quad Notes - Volume 1, A Review of Existing Designs" which takes a fresh new look at the Quad's capabilities.

2-30. SIX METER ANTENNAS

Antennas for six meters are readily available commercially, but also are very easy to homebrew. There are many homebrew antenna project listings on the web – see URL: http://ac6v.com/antprojects.htm

A horizontal dipole for six meters is about 9.36 feet long at the low end of the band. At this length, it is easy to make a rotatable horizontal dipole from aluminum tubing. Or make a center fed vertical at this length, just run the coax feed at 90 degrees from the vertical dipole. For a center fed dipole, the formula is 468/frequency in Megahertz. Cutting the antenna a few inches longer will allow for trimming in order to obtain a 1:1 match. An inverted vee is about 3 to 5% shorter than a dipole at the same frequency.

Verticals are popular, being omni-directional with a low angle of radiation, and having low height requirements. For a quarter wave vertical, the formula is 234/frequency in Megahertz, about 4.68 feet. Radials are the same length. A 5/8 wave vertical is about 11.7 feet high

A three-element Yagi will perform admirably on six meters and makes a nice weekend construction project. For the more exotic propagation modes, such as meteor scatter, more elements are usually needed. Other antennas for six meters include halos, quads, quagi's, J-Poles, squalos, sloops, turnstiles, long wires, and many more.

While antenna polarization makes little difference for DX work, it is a consideration if you also want to work other six-meter operators within ground-wave range.

Most operators with Yagis or rotatable dipoles use horizontal polarization, so if you have a vertical antenna, you may miss out on local and regional six-meter nets, which can provide a nice way of keeping up with weak-signal VHF happenings. On the other hand six meter repeaters are vertically polarized. Cross polarization can be as high as 20 dB.

Many six meter operators have both vertical and horizontal antennas. Be aware that some USA regions favor vertical polarization and others horizontal for ground wave work, ask the local six meter gurus for the regional preference.

2-31. 6M ANTENNA HEIGHT

How high should a 6 meter horizontal antenna be? Several studies indicate that for sporadic E (Es) openings, a height of about 30 feet is optimum. Heights up to 50 feet will lower the radiation angle which is desirable for F layer skip, but above that height, diminishing returns and split lobes may occur. For tropo and other modes, the higher the better.

Verticals can be ground mounted if you have good ground or lots of radials. Elevating the vertical will work well with fewer radials. Some people have multiple antennas at multiple heights to work different kinds of propagation.

The antenna angle of radiation is important for achieving long skip distances.

For a lightly ionized F- layer atmosphere, the high angle radiated signal may go right thru the layers with little returning to earth. But the same signal squirted out at a low angle of radiation may have sufficient "bending" to get back to earth.

Antenna Placements. Often asked is what effect is exhibited when antennas are close to one another ?? One expert advises, measure the antenna impedance with no other antennas around it. Then measure its impedance with other antennas in the vicinity. Any change in impedance indicates mutual coupling. For multiband antennas – this means impedance measurements on each band.

2-32. HOMEBREW ANTENNAS

In addition to the ARRL Antenna Handbook and its compendiums, Bill Orr W6SAI has books on verticals, beams, wire antennas, as well as quads. Highly recommended.

The G5RV is a popular antenna design in so much as it covers several bands, however it may represent a compromise over full size dipoles. G5RV antennas usually require a tuner and the ATU in many rigs only cover a limited range i.e., 3:1, so an external tuner may be required for the various G5RV configurations. Multiband antennas are a neat way to have one antenna for the HF bands; some are resonant on all bands eliminating the need for an antenna tuner. In general a shortened antenna is less efficient than a full size antenna, beware of any claims to the contrary.

There are many homebrew antenna articles on the web such as the St. Louis Vertical, fractual antennas, beverage and loop receiving antennas, rhombics, bazookas, battlecreek special, bobtail, and many others at URL: http://ac6v.com/antprojects.htm

If you have the real estate and some tall trees, try a long wire antenna coupled with a manual or automatic tuner. These are long antenna wavelengths relative to the frequency and will exhibit gain and directionality. See the ARRL Antenna Handbook for several variations which include rhombics, vee beams and others.

The SGC Smartuner family has auto tuners which cover the HF band with power inputs up to 500 Watts PEP. Be aware that a long wire is several wavelengths long, shorter antennas are best described as random wires.

2-33. LOW BAND ANTENNAS

Unless you have some very high supports and lots of acreage for dipoles, 160/80-meter transmitting antennas for Amateurs are usually verticals which will exhibit a fairly low radiation pattern for DXing.

An inverted vee antenna might be used for higher angle radiation for domestic coverage. Other schemes use quarter wave slopers and shunt-fed towers. In addition, you will find it necessary to have a low-noise receiving antenna such as a beverage, or loop antenna.

Both the ARRL Antenna Handbook and ON4UN's book on Low Band DXing are excellent references. Details on several of the above antennas are available on the Internet at URL: http://ac6v.com/antprojects.htm

An inverted L is a favorite since it can be easily installed. NS8O has a design that overcomes the low impedance at the feedpoint. His 160-meter L Antenna is 165 feet long with the vertical section 40 feet high. More info is available at URL: http://www.qsl.net/g3pto/160l.html
A ground system of at least 32 radials is required for efficient performance if it is ground mounted. Some installations use elevated radials, starting the antenna 15' or so above ground and using 4 to 6 elevated 1/4 wave radials.

A 60+-foot tower can be shunt fed. The same ground system requirements are necessary. A 100+ foot tower can be used for a single element or an array by installing verticals from the top set of guy wires. ON4UN's book has details for phasing the arrays.

The four square array uses four verticals located in the corners of a square, each vertical is one-quarter wavelength. Using a phasing network, directivity can be achieved along the diagonals of the square.

The square is oriented so that one diagonal favors Europe and the others will cover other parts of the world. Several schemes can be used for the phasing box. A remote switch box in the shack allows selection of four directions.

The high noise conditions on the lower bands really create a listening problem. Many DXers use a low-noise receiving antenna such as beverages of 500 feet plus, a slinky beverage, or loop antennas. A popular loop scheme is in a diamond configuration – 5 foot to a side using coaxial cable. Loop antennas require a low noise tunable preamplifier such as the Ameco or Palomar series.

2-34. 40 AND 30 METER ANTENNAS

The size of forty and thirty meter antennas are manageable enough to consider rotatable dipoles or two-element beam arrays. Most of the antenna manufacturers offer add-on kits for their Yagi antennas. Also there are rotatable dipoles and in some cases two-element beams available. But these are big antennas and require special consideration of the supporting tower and rotors.

Wire dipoles, verticals, inverted vees, slopers, bobtails and half square antennas are some of the alternatives.

Many opt for the 40-meter vertical or phased vertical arrays. With a quarter wave being 32 feet, it is feasible to build a vertical with aluminum tubing or fiberglass poles. The author constructed a 20-foot fishing pole coupled to a swimming pool pole (total 32 feet) so the radiator feed point is about 5 foot from ground and uses two elevated radials.

Phased arrays result in gain and directivity and details can be found in the ARRL Antenna Handbook and ON4UN's book on Low Band DXing.

Bobtails are vertical arrays that look like the letter "E" turned 90° clockwise onto its tails. They have three-quarter wave vertical elements and are one wavelength long.

The Half Square is a shortened Bobtail and has two-quarter wave vertical elements and is a half wave long. The tops of the verticals are all connected together. The bottoms of the verticals are all insulated from ground. Bobtails are fed from the bottom of the middle element. Half squares may be fed at either end.

The G5RV antenna is very popular for the 40 through 10-meter band and its original configuration requires a tuner for proper operation.
The antenna is a 102 feet long, center-fed design. W6RCA's Geocities Web Page discusses the means to operate without a tuner. These can be homebrewed or are available commercially. For good 40-meter operation, dipoles, G5RV antennas, and other horizontal antennas need to be up in the air for good low-angle radiation, preferably a half-wave length or 64 feet at 40 meters.

The Saint Louis Vertical is another popular antenna for the 40 through 10-meter band. Details are at URL: http://www.amqrp.org/projects/projects.html

The Battle Creek Special is a full-sized 1/4 wave vertical on 40 m, a loaded 1/4 wave vertical on 80 m and a loaded 1/4 wave inverted L on 160 m. The feed-line is a single 50-ohm coaxial cable and no tuner is required for 160 m, 40 m CW and the 80 m CW bands.
The tricky part is the trap design, some using coaxial traps. Details can be found at OK1RR's site at URL: http://www.ok1rr.com/

2-35. STEALTH ANTENNAS

Twenty years ago nearly everyone had antennas on their rooftops to receive TV signals, but today all across country, Homeowners Associations and local restrictions are forcing amateurs to go into the stealth antenna mode.

Verticals in trees, hamsticks, screwdriver antennas, mobile installations, isoloops, small gauge wire and other devious means such as up at night – down at dawn installations are used. Several schemes can be found at URL: http://ac6v.com/antprojects.htm#STEALTH

Also you may want to read antenna reviews on HF Directional Antennas, HF Mobile Antennas HF Vertical and Wire Antennas at URL: http://www.eham.net/reviews/

For example, Isoloops are very high Q (narrow bandwidth) and need constant retuning. Hamsticks are shortened low efficiency antennas but RF gotta go somewhere and you do what you gotta do for stealth.

2-36. ANTENNA GAIN

Antenna gain is somewhat a misnomer, as "Ya can't get out more than ya put in". In reality no antenna has gain, only losses due to inefficiencies So what is meant by antenna "gain"? The gain spoken of has to do with shaping the antenna radiation pattern so that more energy is radiated along a given plane as compared to a reference antenna. In effect the radiation pattern is squished into a narrower directional beam. Squeeze a round toy balloon to visualize the effect.

Several reference antennas can be used, but the ones most often used are the dipole and an isotropic antenna. When using the dipole as a reference, the subjective value of zero dB gain (dBd) is assigned (except NMEA, National Marine Electronics Association, which assigns it a subjective value of 3 dB gain). The isotropic antenna is by theory an infinitely small sphere that radiates equally in all directions. By comparing a real antenna to this standard, a figure of merit can be expressed in dB, usually stated as dBi. That is, the gain of the real antenna over the isotropic reference. Since the real antenna does not radiate equally in all directions, some plane of directivity must be specified.

Every antenna, other than an isotropic will have gain in some direction at the expense of that in others. This gain of course, comes from not radiating power in all the other directions. For HF beam antennas, the beam width might be in the order of 30 to 60 degrees, it is not the 1-degree some might hope for. And the beam width is measured at the minus 3 dB points (half power points) so the antenna "hears" and "talks" well beyond the beam width specification. Dipole beamwidths can be much as 130 degrees depending on height above ground.

2-37. TOWERS AND ROTORS

Several categories of towers are available from rooftop tripods to the sky is the limit. These include crank up masts, crank up towers, compact crank up towers, heavy duty crank up towers, fixed towers, guyed, free standing, and bracketed. Local regulations can drive you goofy with all kind of forms, engineering drawings, etc.
Best consult with a tower owner in your area before anything else. Then start the laborious process of checking the local regulations and the city engineering department for permits. Also the tower manufacturers will supply tons of information regarding installation. Tower installation can be complex as well as dangerous, so best have experienced or professional help. See Champion Radio Tech Notes for tower tips URL: http://www.championradio.com/

Rotors. Rotors are rated as to wind load, turning power, brake power, brake type, and bearing type. Needless to say, if buying used, don't rely on the previous owner to have selected the right model. Check the rotor specs in view of the antenna to be turned - favoring the heavy-duty models if you expect to do a lot of contesting or rapid fire DXing.

Use a high quality thrust bearing above the rotator. Rotors can suffer greatly during a contest - they may do more work in one contest than in several months of normal operation. Rotors which cover more than 360 degrees by allowing the antenna to move past the normal end-stops can be a real time saver.

With some rotors, operators may tend to release the brake switch before the antenna has stopped moving – resulting in possible damage to the wedge brake. An electronic mod fitted to the control box will perform the proper sequencing. Maintenance of rotors and coax is a necessity every few years. Check the coax for excessive loss and lubricate the rotor as per the manufacturers manual. If you wait until it fails – Murphy's Law says it will fail during a contest or while swinging the beam to North Korea!!!

2-38. BEARING AND DISTANCE

Since some antennas have forward gain (Yagis, Quads, etc.), a rotor is required to point the antenna in the desired direction. Or several fixed antennas along desired paths. HF directional antennas have a fairly wide beam width (measured at the half power points) perhaps 50 degrees or so. So exact headings are not required and the antenna "hears and talks" well off the half power points. Usually the heading is the short path along the great circle bearing between you and the DX, but long paths, skewed paths and gray line paths may also exist.

Long path is somewhat the reciprocal heading but may vary, due to gray line or skewed paths. Programs are available to provide beam headings and distance both short path and long path – see Chapter 3. In these, you enter in your longitude and latitude and the program gives the beam headings and distance.

Some take into account the MUF and predict the probability of a propagation path along with number of hops under the present conditions. Also there are on-line lookups for this information.

2-39. GROUNDING AND COMMON MODE

An excellent source for Ham Systems grounding is at URL: http://www.polyphaser.com/ And http://www.geocities.com/SiliconValley/2775/gndsys.html These are a must read.

The common mode and balance debate is often a subtle point for many beginners. Coaxial cables are physically asymmetrical with respect to their external physics but are symmetrical with respect to their internal physics, and so if all currents flowing on the coaxial conductors can be successfully confined to their insides they can be operated as a balanced system.
Coaxial systems are considered balanced when there is no common mode currents flowing on them; that is, all of the currents are flowing inside the coax and not on the outside surface of the outer shield conductor.

Stated differently, with coaxial systems, balanced simply means that the current in the center conductor of the coaxial cable is exactly matched by the current flowing on the inside of the outer conductor in the opposite direction. Any current flowing on the outside of the outer conductor is then an undesirable common mode current and causes unbalanced operation.
Common mode current chokes, often called coaxial baluns, can choke off these outside outer conductor currents and force a balanced operation. Grounding usually adds little to achieve a balanced system.

2-40. RFI SAFETY AND RFI TO THE NEIGHBORS

When submitting some of the FCC forms, it is required to perform an RF safety check for near field radiation. This can be found on the web at URL: http://www.arrl.org/tis/info/rfexpose.html

Working all states is one thing, but WAN (worked all neighbors) is quite another and big antennas and running power, even 100 Watts can result in a whole new set of problems. An excellent document to read is the FCC INTERFERENCE HANDBOOK at URL: http://www.fcc.gov/cib/Publications/tvibook.html

Proper grounding of equipment may solve some RFI problems, see PolyPhaser document at http://www.polyphaser.com/ There are several manufacturers that make kilowatt band pass filters for the output of your transmitter to reduce harmonic radiation. These typically offer up to 60 dB of attenuation. Placing two low pass filters in series, will in theory, yield 120dB, but in practice, it is more like 90 or 100dB. At these levels of attenuation, signals tend to go around the filters. Radio Shack and others offer input filters for TV and FM sets. RF Parts offer RFI kits including toroids and chokes. For stereo, a 0.01 uF cap across speaker leads will bypass speaker coils. Be sure neighbors have high quality coax – both cable and outdoor antenna lead-ins. Replace old twin-lead TV antenna lead-ins with coax and transformers. Clean up TV antenna connections.

Be proactive in solving the interference issue. (Caution: don't open any neighbor's equipment!) Provide the filters free and show them how they are installed at YOUR location and recommend they contact a service technician. Having a neighbor actually see that your house is interference free is a strong indicator that their house can be likewise, but with some additional effort. For telephone interference, see the FCC Telephone Interference Bulletin CIB-10 August 1995 at URL: http://www.fcc.gov/cib/Publications/phone.html K-Com Company offers telephone filters which are very effective for reducing telephone interference.

2-41. LIGHTNING PROTECTION AND GROUNDING

Consider carefully how to protect your equipment and your home from lightning. Even in areas where lightning storms are less common, there are usually a limited number of storms each year. Take the time to study the best way to protect your installation from the effects of lightning by consulting reference material on the subject.
This subject is covered quite nicely in an 11-page discussion at PolyPhaser at URL: http://www.polyphaser.com/ Be sure to read this if you are in an area that experiences lightning strikes. They also have a nice write-up on station grounding.

The installation of a lightning arrestor is a start, but there is more that you can do. For example, terminate your antenna system transmission lines at an entry panel that you install outside your home. Ground this entry panel to a good outside ground, and then connect appropriate feed lines between the entry panel and your transceiver. When a lightning storm occurs, you can ensure added protection by disconnecting the feed lines from your transceiver.

CAUTION. DO NOT ATTEMPT TO USE A GAS PIPE OR AN ELECTRICAL CONDUIT (WHICH HAS THE WHOLE HOUSE WIRING ATTACHED AND MAY ACT LIKE AN ANTENNA), OR A PLASTIC WATER PIPE FOR A GROUND.

At a minimum, a good DC ground is required to prevent electric shock. For superior communications results, a good RF ground is required, against which the antenna system can operate. Both of these conditions can be met by providing a good earth ground for your station. Bury one or more ground rods, or a large copper plate under the ground, and connect this to the transceiver GND terminal. Use heavy gauge wire or a copper strap, cut as short as possible.

One DXer I know rented a hydraulic drill and went way down in the earth in several places and made an extensive ground system. He experienced a marked reduction in noise levels; however, your mileage may vary.

2-42. COMPUTER EQUIPMENT

Computers are more and more a part of the average Amateur Radio station. Not only do they provide rig control and decoding, but offer a wide variety of programs for operating and allow using many of the operating aids available on the Internet. Many of the older computers will work fine for DX packet clusters, logging, etc., depending on how extensively you load in programs. Be aware that some programs such as PSK may require processors of 100 MHz or higher and Windows 95 or better and a compatible sound card. Some of the DXing aids that a computer will give you are; rig control, rotor control, prefix finding, beam headings, great circle programs, distance and bearing, satellite tracking, DX Packet Cluster Telnet, NCDXF/IARU Beacons, antenna modeling, propagation prediction, logging programs, Morse and pileup practice, maps, grayline mapping, DX and contest programs, packet, and using a sound card - several data programs such as PSK31, MFSK16, RTTY, CW, and SSTV. See Chapter 3.

A Ham computer can be a great aid for the DXer, but also a real nuisance noise generator. There are two FCC designated computer classes – A (home computers) and B (industrial strength). The class B computers supposedly conform to higher standards of radiation emissions and a used industrial-strength computer may be a preferred choice. But the best bet is to test the computer alongside your HF equipment or take an SWL portable with you when looking for a Ham computer.

If you are considering a new purchase of a computer for your Ham Shack, consider a Laptop, as these are generally much less noisy than the full size computers.
Laptop computers with their non-CRT screens usually have a lower level of emission than the CRT types. In addition there is an opportunity for the manufacturer to tightly shield the laptop, some do - some do not. Test them in your environment.

The author had a well-known desktop computer that emitted birdies, noise, and trash all over the HF bands; it could totally trash weak DX signals. After a lot of shielding, ferrite beads, etc., it quieted down quite a bit. Then I borrowed my XYL's Toshiba Laptop and almost all the noise disappeared!! Your mileage may vary. Caveat Emptor. If you intend to get into the new digital modes, e.g., PSK31 etc., make sure the computer sound card is compatible.
In contesting setups, it is typical to have toroids on virtually EVERY cable connected to the computer and audio isolation transformers on audio cables connecting the computer soundcard and the rig.

Things to look for when evaluating a new computer are: chassis lips fit together tightly with the cover, use of finger stock RFI shielding (available at computer part stores), and multiple cover screws for a tight fit. The power supply should have good AC line filtering. Computer noise is usually stronger on VHF. Use a 2-meter HT, set the squelch on the edge and then hold it close to the monitor and computer.

Turn your computer on, switch your receiver to a shielded dummy load or simply disconnect the antenna, and tune across the Ham spectrum listening for "birdies" or other noises. For HF, check the middle of the HF spectrum, 10 - 20 MHz, as it seems the worst affected.
Note the frequencies and the S-meter readings. Next, turn off the computer and go back to the noted frequencies and check for differences. Here are some tips for quieting your computer.

Disconnect one cable at a time from the computer while watching for changes in noise levels. Often a particular cable acts like a noise generating antenna.

Try a snap-on ferrite toroid attached to the cable where it exits the computer. It may require 3 or 4 toroids on a stubborn cable. Up to 30 dB per toroid may be realized. Redoing the cable in shielded wire will also help.

After disconnecting all cables and there is still noise - then the noise is either being transmitted via the power cord or coming out of the CPU case itself.

The same toroids and other line filters can keep the noise off the AC line. If the noise is leaking out of the CPU case, it calls for an experienced technician to open up and work inside the case.

WARNING: HIGH VOLTAGES ARE PRESENT INSIDE THE COMPUTER CASE AND IF CONTACTED CAN RESULT IN INJURY OR DEATH, BE SURE YOU KNOW WHAT YOU ARE DOING AROUND HIGH VOLTAGES.

The main things to check are case shielding and board grounding. Often with metal cases, some paint scraping and star-lock washers will help. Spray-on conductive coatings and plastic-sheathed foil panels are commercially available for plastic case improvement. If you already have a noisy computer and want to quiet it down, an excellent article by 9V1ZV, Daniel Wee Chun Chian is available at http://ac6v.com/comprfi.htm

2-43. COAX CABLE AND CABLE MANAGEMENT

Coaxial Cable. DX and contest stations usually use coax as opposed to open wire feeders which have very low loss. Open-wire feeders require an antenna tuning unit to match the 50-ohm unbalanced output of the transmitter to the 600 ohm balanced open-wire feeders. It is difficult to construct an antenna tuning unit which can handle the power levels used at contest and DXer's stations. Open-wire feeders are also mechanically tricky, and need to be separated from other feeders and metal objects to perform properly.

You should be concerned with coax cable losses. Use the best coax that you can afford paying attention to the dB loss per 100 feet at the frequencies of operation. It's easy to test for losses and it should be done periodically to keep your station in top operating condition. Keep a note of the results and you can quickly determine if water has seeped into the coax or it has increased losses.

Use a power meter, 50W dummy load, and a low power transmitter – 30 MHz is frequently specified in cable charts. Connect the 50W dummy load to the antenna end of the coax. Then adjust the transmitter for a measured 10W into the coax at the transmitter side, then move the power meter to the antenna end near the dummy load and measure the power arriving there.

At low radio frequencies, most coax loss is due to conductor resistance. At high frequencies, much of the loss is due to the dielectric material.
Characteristic impedance (Zo) is not a function of frequency over a wide useful range. Zo can be the same for cables of widely varying external size. Larger cables of the same Zo have less loss and more power capability in most cases.

Typical results for good coax are:

RF Input power to coax cable = 10W at 30 MHz
RF Output power from Heliax 100 feet of LDF4-50 = 9W
RF Output power from coax cable 100 feet of RG213/U = 7W
RF Output power from coax cable 100 feet of RG58A/U = 5W

Well now - does that make a good case for using the best coax??? And Heliax offers a solid outer conductor instead of a percentage of "braid" like RG213 with 90 or 95% braid. But heliax is hard to handle and install, nevertheless, it is popular with the big guns. Also be aware of the coax power ratings. RG58 cable is rated at 300 to 400 watts at 30 MHz.

But also be aware that coax power handling capabilities should be de-rated as the SWR goes up. Coaxial cable specs can be found at URL: http://www.cablexperts.com/

Cable Management. To accommodate the myriad of cables in a Ham shack having 3 or 4 receivers, speakers, TNC, etc., one can end up with a snake basket full of wires and cables. One solution is to use cable trays made by Panduit at URL: http://www.panduit.com/   Best use separate trays for audio and RF lines. Another is to use electrical conduit - square or round.

Another idea is to visit the Hi-Fi stores who sell split plastic conduit for bundling hi-fi cables. To prevent RF from getting into things, it is advisable to use shielded wire on audio as well as RF leads. If you are running power, best use the larger coax types.
However, these can be quite stiff, making bends difficult. Several cable vendors sell a "Soft-Flex" version which makes coax installation in the shack much easier.

For feeding cables into the house, folks have used vent pipes thru the house roof, dryer vent tubes into an exterior wall, and some construct plastic or metal boxes as feed-throughs complete with coax connectors on the exterior and interior of the house wall. Seal the ends to prevent bugs and critters from gaining an access point. Radio Shack used to sell a plastic wall feedthru with seals on either end.

2-44. STANDING WAVE RATIO

For this topic it is not our intention to go into a long discourse on SWR. It is a highly complex subject on which whole books have been written and need not be repeated here and some gurus have argued for years without agreement. If you do nothing else regarding SWR be sure to read the better written articles in books and on the ARRL website. Namely:

ARRL Handbook http://www.arrl.org/catalog/

ARRL Antenna Handbook http://www.arrl.org/catalog/8047/

Reference Data For Engineers: Radio, Electronics,Computers, and Communications" Seventh Edition, Edward C Jordan, Editor in Chief, Published by Howard Sams & Co. ISBN# 0 672 21563-2

Perhaps an easy way to visualize standing waves is to consider a bathtub with a level of water in it. Dropping an object in the water will send a wave outward to the edges.

If one end of the tub has a totally absorbent sponge-like substance, then the sponge will absorb the waves as they impinge. On the other end, consider a mirror-like surface that reflects the incident wave. As the forward wave encounters the reflected wave, adding and subtracting takes place and a standing wave pattern is developed.

SWR formulas are E_{max} divided by E_{min} OR if the load contains no reactance, then it is the ratio of the load resistance and the characteristic impedance of the line. R divided by Z_o OR Zo divided by R if the R is less than Z_o (load with no reactance).

SWR is a well-worn subject often misunderstood or surrounded by myth.

High SWR does not mean an antenna cannot radiate a decent signal, nor does it necessarily mean you will have RFI (TVI) problems, nor does it imply excessive radiation from a transmission line. And a low SWR does not necessarily mean the antenna system is working properly, it may have excessive losses due to faulty cable and still give an acceptable SWR with little power left for the antenna. A 1 to 1 match into a Heath dummy load won't get RF into DX land – hi hi.

Before the days of coaxial cable, open wire was (and still is) used to obtain very wide bandwidths with high efficiency.
This is because all power reflected from the line/antenna mismatch is conserved, not dissipated and is returned to the antenna by an antenna tuner. There is some loss due to SWR but the additional loss is usually negligible due to the low attenuation of open wire lines.

Coaxial cable however can exhibit losses which **may** be of concern because the permissible reflection and SWR limits are lower than in open wire. When the coax line is very long or lossy, the increased loss due to high SWR can be calculated by referring to the ARRL pages by W1GV URL: http://www.arrl.org/tis/info/pdf/7901019.pdf

But also be aware that coax power handling capabilities should be de-rated as the SWR goes up. Power handling capacity in watts is inversely proportional to SWR. Using the larger coax types and full legal limit power, SWR's of up to 3:1 should not present a problem. For quality coax and reasonable lengths (100 feet or so) by using tuners (internal or external to the transceiver), nearly all the power will be delivered to the antenna, with very small attenuation due to SWR on line. Running high power however, with high SWR can result in damage to the feed line, filters, or antenna tuner. See COAXRATE at URL: http://www.smeter.net/feeding/coaxrate.php

Both the ARRL Antenna Handbook and Reference Data For Sams, Engineers: Radio, Electronics, Computers, and Communications state that the power rating for lines are for unity voltage standing-wave ratio. Safe operating power is inversely proportional to SWR expressed as a numerical ratio greater than unity. Note: This is a safe operating area of coax cable (break down voltage) not to be confused with loss factor in coax as a function of SWR, which is covered in many text books.

The experts will tell you that any effort to reduce an SWR of 2:1 on coaxial line will be non-productive from the standpoint of increasing power transfer significantly. The DX receiver on the other end could not detect the difference.

Adjust the antenna for lowest SWR in the part of the band you operate in or the center of the band and don't get overly paranoid about SWR as many hams do. Here again – refer to the books as cited above.

When working with VHF and UHF equipment, losses can be appreciable and it is almost mandatory to match the system and use low loss coax.

SWR does cause a problem in modern transceivers with solid-state outputs. Solid-state transceivers are designed to detect a poor SWR and reduce the output power in order to achieve a safe level. An SWR of 2:1 or less is generally tolerable causing no harm to the transceiver. Antenna tuning units (ATU's) can be used to match impedances but do not remove or reduce a poor SWR. Generally, Vacuum tube finals are much much more tolerant of high SWR conditions.

DO BE concerned if your antenna suddenly exhibits a marked change – calls for a system checkout. Also the SWR gurus will advise making your SWR readings at the antenna, not the transmitter which most of us do as a matter of convenience.

Keep in mind that a good match only means that the transmission efficiency of the line is good but says nothing about the radiation efficiency of the antenna which may have excessive losses from several effects – poor ground (worm heating), coax losses, traps, etc.

For example, the VSWR can be out of wack due to a poor ground or having the feed line in the field of the antenna thereby inducing unwanted currents into the feedline. Here a choke or balun should be used when matching a balanced system to an unbalanced feedline. Low VSWR has nothing to do with the radiation angle which may be poor for your application.

Finally, be absolutely sure to read the reference documents above as it dispels the myths surrounding SWR and will keep you from running up and down the tower trying for the perfect system which in many cases is not worth the effort.

2-45. ANTENNA TUNERS

An antenna tuner is an impedance transformer, sometimes called a transmatch. The function of the tuner is to match impedances in the system. They do not get rid of VSWR and if you don't believe that, insert an SWR meter between the tuner and the antenna, not between the transceiver and the tuner. The function of the tuner is to match impedances from the 50-Ohm output of the transceiver to higher or lower impedance presented by the antenna.

Thus the tuner input is 50 Ohms and the transceiver output is 50 Ohms and the transceiver is happy. When using an adjusted tuner, the system is not mismatched and does not suffer from a mismatch loss. Always use the lowest possible value of inductance that will give a proper match. Using higher inductance can result in wasted power inside the tuner, and may cause damage. Don't try to match extremely low or high impedances. At best, tuners can squander 1or 2 db. Maybe build your antenna non-resonant and match it with transmission line as the transformer.

There are many tuners available but when you choose a smaller model (100W to 300W) be aware that the low power tuners may have much higher losses than the big tuners depending on the size and quality of the components. And serious mismatches can result in high voltages across the tuner components resulting in failure. For example when operating at VSWR's of 10:1, the losses can be appreciable in some of the smaller units. Also you might want to get a linear some day and it might make sense to get the kW model to start with.

Using a manual tuner in the DX environment can be a pain because as you change bands or even tune across a wide band, you will have to readjust the tuner for best SWR. Having antennas that are self-resonant for your operating bands are a convenience. But there are lots of antennas that work well with a tuner and open wire transmission lines – see the ARRL Handbook. And a linear that does not require retuning when switching bands is a contesters/DXers dream. There are automatic external tuners available see URL: http://ac6v.com/antdealer.htm#TUNER

Most internal tuners (ATU) are not much more than glorified "line flatteners". They were never intended to match complex antenna systems. They are intended to be used with antennas that are tuned. The main purpose of an ATU is to provide a wider usable bandwidth for the radios finals when using existing antennas. Example: say you had a dipole that covered 3.7 to 3.9 MHz. The tuner will let you load to 3.55 MHz and keep the radio happy. And the losses are still fairly low. Most internal ATU's were never intended to run "all band" antennas, where the Z can be from very low to very high. For that, you need a outboard tuner.

2-46. MICROPHONES

If you listen carefully to a pileup, the signals that seem to stand out are the YL's, the kids, and audio that has a fair degree of crisp high notes. Tailoring your audio can be a real advantage in the rumble of a pileup. May not beat out the big guns, but it sure helps in the long run.

According to research on the intelligibility of speech in hearing English words, the frequencies important for speech intelligibility are the consonant sounds from 500 to 4000 Hz. They contribute 83% of word intelligibility.

Frequencies from 500 to 1000 Hz contributes 35% of word intelligibility and 35% of sound energy. Frequencies from 1000 to 4000 Hz contribute 48% of intelligibility but has only 4% of sound energy! In contrast, frequencies from 125 to 500 Hz contributes 55% of sound energy but only 4% to word intelligibility.

The bottom line is nearly half the speech intelligibility is contained in 1000 to 4000 Hz frequency range with only 4% of the speech sound energy. On the other hand, the low frequencies 125 to 500 Hz have most of the speech energy but contribute very little to intelligibility.

The Astatic D-104's are famous for their high-frequency characteristics, rising to about 10 dB at 3,000 Hz. For today's solid-state radios, an amplified D-104 is recommended as it provides a proper impedance match.

Bob Heil has perfected the art of audio for Amateur Radio and offers a line of microphones for DXing or rag chewing or both (HC-4 and HC-5 elements). A great addition for DXers/Contesters is a combination headset boom mic and a footswitch, frees both hands for typing and logging. W2IHY has a variety of microphone equalizers and an on-line demo of enhanced DX audio – a must see and hear at -- http://w2ihy.com/

The new digital radios (Kenwood TS-870) offer built-in tailoring for transmit audio. This feature is very powerful for adjusting your audio to suit your voice and working conditions.
Features include provision to adjust the Processor low and high frequency response, Transmit Bandwidth and Transmit Equalizer.

Although the effects of audio can be heard in a headset using the Monitor function, a much better method is to monitor your audio with a nearby receiver and a good quality headset, then you'll be sure of what your audio sounds like on the air.

2-47. HEADPHONES AND SPEAKERS

It is tempting to buy a matching speaker with your rig, but most weak signal DXing is better accomplished with a pair of headphones.

Speakers are OK when working strong signals or rag chewing and some have built-in filtering to cut highs or lows or both.

At least one manufacturer has provision for a popular DSP unit which can be operated in the speaker mode or headphones can be plugged in to take the full benefit of the DSP unit.

Over the years, the standard for headsets has been those tailored for the Ham audio spectrum (300 to 3,000 HZ). If you have perfect hearing, these may well be your choice.
As one's hearing falls off with age, the use of full range headphones may be more effective. For SSB, the high-frequency notes are important for intelligibility and a $20 pair of titanium phones may work well for you.

Try a variety of communication phones and HI-Fi phones for comfort and a good match to your hearing. These are also excellent filters for screaming spouses and harmonics and it allows you to work CW while the family is sleeping.

The author has two sets of headphones; one is the cover-the-ear type when the house is noisy. The other is a lightweight sit-on-the-ear type for long use such as contests. For some, the cover-the-ear type may get hot and sweaty after long use. Heil Pro and Sennheiser phones are favorites.

As mentioned under microphones, nearly half the speech intelligibility is contained in 1000 to 4000 Hz frequency range with only 4% of the speech sound energy.
On the other hand, the low frequencies 125 to 500 Hz have most of the speech energy but contribute very little to intelligibility.

The intelligence in an audio signal is contained approximately from 750 to 4000 Hz and you want to be sure you can hear this range either by using non-communication headphones or boosting or equalizing as necessary.

High frequency tones crack thru a pileup – the YL's, kids, equalized microphones (like Bob Heil's) are the ones heard thru the rumble. Consonants (2000 Hz and up) like S and T, seems to pierce through.

Replacing communication phones (300 to 3,000 Hz) with Hi-Fi phones can make a difference for older operators or those with a reduced hearing range. A great addition for DXers/Contesters is a combination headset boom mic and a footswitch, frees both hands for typing and logging.

Also, MFJ sells an MFJ –616 Speech Intelligibility Enhancer which works well for tailoring your receive audio, just might give you that DX edge. A common hearing loss occurs at the 4K-octave range. Boosting this range adds considerably to intelligibility.

A favorite with some operators is to wire the two phones out of phase or use a switch to select either in phase or out of phase headphone operation. For some this makes copying high speed and low signal strength CW much easier. For more see "Hearing CW in Noise by Chuck MacCluer, W8MQV". – see URL: www.nitehawk.com/rasmit/br_cpy.html

Some operators use stereo phones with a variety of filters in each earpiece to create a pseudo stereo effect. This uses a low cut filter in one earphone and a high cut filter in the other without phase reversal.

Use a low-pass filter to one ear and a high-pass response to the other with the corner frequency of both filters being equal, e.g. 1000Hz. Then 1000Hz signals seem to be in the middle of your head with higher frequencies on one side and lower frequencies on the other.

2-48. ESP HEARING

Perhaps "Expected Sensory Perception". Why can some operators pull signals out of the noise and others fail?? Some long time DXers claim the best filter is between the ears. Practice listening to signals in the noise, you will begin to catch syllables.

The brain apparently begins to filter out the noise and you can hear a lot more than you can imagine after some practice. Tune to an apparent blank frequency and listen intently and you sometimes can hear signals way down in the noise. Remember that there is no person-made noise or power line noise on a DXpedition Island!!

If you can discern the DX then CALL EM – you will be surprised how many will come back. Some expert CW operators can pull out signals that are right at the noise level. A new product by AmCom is "ClearSpeech" claiming a noise reduction of 12 dB on all types of noise.

Their e-mail address is: amcom@digisys.net Several noted DXers have evaluated this unit with excellent results – see World Radio – July, 2001 issue.

2-49. KEYS AND KEYERS

To send high-speed code you will need an electronic keyer, which is included in most modern transceivers. The key can be single paddle (press one way get dits, the other way dahs) or an iambic type. With the iambic type, squeezing both keys alternates between dits and dahs - or dahs and dits depending on which paddle you hit first.

There are two iambic modes - A and B. The original Curtis chip is mode A, while the Accu-keyer types are mode B. Mode A seems to be the most popular, but if you are just learning iambic, either one will do. To see how these work, send the letter C dah dit dah dit.

The A mode completes the element being sent when the paddles are released, so you release the paddles after you hear the second dit.

The B mode sends an additional element opposite to the one being sent when the paddles are released. So with mode B, you would release the paddles when hearing the second dah. More on iambic keyers and how to adjust paddles in Appendix A1.

Paddle keys might seem expensive (from $50 to $300) but keep in mind you are paying for precisely machined parts, exact adjustments that stay locked, bearings on pivots that allow no side play and contacts made of silver or gold plated to prevent tarnishing.

Ergonomics enter into the design for ease of use during long periods of operating and have weighted bases to minimize movement. In addition the looks of the key enter into the price when they are chrome or gold plated.

Operators have their favorites, but keep in mind their posture, being heavy or light-fisted, all play a part in preferences.

If possible borrow a friends paddles and practice, then borrow another, or go to the candy store. The choice is highly personal, so beware of recommendations until you can try them out.

A memory keyer is a must-have if you get into contesting and a nice accessory if you work DX in the CW mode.

For a rare DXpedition, you can send for hours before you get through and might as well have a memory keyer to do the work for you. For contests the memory keyer can send sequential serial numbers or memorize the reporting system.

If you are the nervous type and don't want to goof a reply, a memory keyer will do all the work for you, just be prepared to recognize your call at any speed and push buttons. Be sure however you know that the DX is calling your district and country and if he/she is working split, usually signaled by DX11DX UP.

2-50. DSP UNITS

Digital Signal Processing has made its way into Amateur Radio. Many new rigs have DSP built-in and the technology has reached the point where they are effective on certain types of noise. For older transceivers there are outboard units available such as the Time Wave units.
In addition to noise reduction, these will really give good selectivity as well razor sharp filtering on CW signals. However don't expect miracles, particularly on very weak signals.

A new product by AmCom is "ClearSpeech" claiming a noise reduction of 12 dB on all types of noise. Available as a Mobile or Base configuration. Their e-mail address is: amcom@digisys.net Several noted DXers have evaluated this unit with excellent results – see World Radio – July, 2001 issue.

Beware of the units that are inserted into the antenna line – these may cause a loss in signal strength – test with and without the device

2-51. ANTENNA ANALYZERS

MFJ, Autek, AEA and other antenna analyzers have been on the market for quite a while now and have somewhat replaced the grid dip oscillators for antenna measurements. If you are serious about HF antenna experimenting and building, these are well worthwhile to purchase.

Unlike ordinary SWR meters, the antenna analyzers allow you to quickly determine the antenna resonance, SWR and impedance characteristics. Additionally, the serious Ham builder can use the expanded measurement capabilities for analysis and measurement at the bench.

They can be very handy for checking RF components and resonant circuits as well as antenna, feed line and antenna coupler adjustments.

Features of some of the these units include: impedance measurements including RF chokes, baluns, stubs etc., SWR, bandwidth measurements, L&C measurements, resonant frequency of traps, converting between L and C and Z, measure 1/4 and 1/2 wavelength transmission lines, determine cable loss and impedance.

They can accurately check the resonant frequency of an HF antenna system, from 1.8 MHz to 30 MHz. They also provide a fairly accurate signal generator for the HF bands. VHF models are also available.

2-52. SWR AND POWER METERS.

In addition to an antenna analyzer, almost all DXers use an outboard Power and SWR meter permanently connected in the transmission line. The standard has long been Bird Wattmeters which are quite accurate, but somewhat expensive. Bird meters also have slugs for many power and frequency ranges for accuracy and convenience in reading. Also the Bird meters yield forward and reverse power readings.

The cheaper power/swr meter are bridge types that detect the voltage in the transmission line and have a meter calibrated accordingly. This is satisfactory over a limited frequency range, so the Bird meters use slugs for each frequency range resulting in increased accuracy.

Some of the less expensive meters are fine for relative indications but have trouble being accurate over a wide frequency range. The cheaper Wattmeter/SWR meters are usually easier to use and allow for rapid tune-ups, however they typically give meaningful power measurements only when the SWR is low.

Many hams are confused when the SWR is high and the power meter reads a higher power level than what the transmitter is putting out. Here the complex waves of combined forward and reverse power are at work giving such readings.

The cross needle types display power and SWR and are very convenient in operation. Also some of the newer rigs have a built in SWR meter which works in conjunction with the automatic tuners. Some manufacturers offer digital SWR/Power meters which although expensive offer exceptional accuracy and an easily read indications.

For very accurate power measurements, the commercial grade power meters such as the Hewlett-Packard types are recommended. Some of these use thermocouples (heat measuring) rather than a voltage detector. Calorimeter Wattmeters measure the heat rise due to RF energy and the calibration is independent of frequency. Others use directional couplers – see ARRL Antenna Handbook.

- SWR indicators need not be placed at the feedline-antenna junction to obtain a more accurate measurement.

- The accuracy limits of the common SWR meters indicate that SWR at any point in the antenna system may be determined by simple calculation involving the SWR at the point of measurement, the transmission line attenuation per unit length, and the distance from the measured point to the desired point.

- If the SWR readings change significantly when moving the SWR meter a few feet one way or the other, it indicates that some other problem exists and not that the SWR is varying with line length. The SWR bridge need not be placed at half-wave intervals to obtain a correct reading.

2-53. PHONE PATCHES

Sooner or later a DX station will ask you to run a phone patch, indeed it may be a way to get a new DX country if you hear a CQ for a phone patch into your area. Be aware of the third party agreements with the various countries so you know it is legal to conduct the phone patch. Before conducting a phone patch, be sure to look up the call of the station requesting the patch, checking on proper name, location, etc. Unlicensed operators (especially maritime mobiles) will assume a call and it is up to you to verify as best you can that the call is legitimate.

Also beware of the conversation that gets into the veiled mode where the parties are talking in a sort of a veiled code. If you don't have a phone patch, acoustic coupling works well; just hold the telephone to the speaker – then to your microphone.

2-54. SCRATCH PADS

In addition to your logbook, use a notebook to write down all calls and info as you hear them whether they go in the log or not. Later you may work them and your scratch pad may yield info not gleaned during the contact (QSL info, ops name, etc.).

Also a DX buddy may ask if you have the info on a particular DX. The packet cluster may also reveal past data, but perhaps no one spotted it.

2-55. MONITOR SCOPES

Some of the newer transceivers have a built in spectrum scope which is handy for spotting band activity near and around your operating frequency. Some are expandable to check the whole band and can reveal pileup activity as well as showing signal strengths across the band.

Use of color is included to alert the user of new activity. Some older rigs had provision for a separate monitor scope including provision to monitor the transmitter outputs including kilowatt amplifiers. These are great for adjusting mic gain and avoiding flat topping of the transmitted signal. Some offer two-tone testing to determine distortion products. Also some monitors can be configured as an RTTY indicator using an X-Y axis presentation for exact tuning.

2-56. FOOT SWITCHES

A valuable accessory is a foot switch that will free both hands for computer entries or hand writing in logs. These can be used with a standard desk or hand mic or better, in combination with a headset and boom mic.

The switch goes across the PTT line available on most rigs at the accessory jack. Be especially careful when soldering up a plug for accessory operation, any shorts on the pins can result in grievous damage to the rig. Use an Ohmmeter to verify integrity of your connections. Most microphone elements are not switched on and off with the PTT switch to allow for VOX or foot switch operation. For SSB, you will love the foot switch during a contest when you are the callor.

This allows you to type on the computer or hand log while giving the reports. Foot switch wander around? See Pedal Sta I or II
See URL: http://www.pedal-sta.com/pedal-sta.html

2-57. OTHER GOODIES

As described in Chapter 9, Logging and QSLing, an accurate UTC clock is an absolute necessity. Even the ones that sync off of 60 Hertz are very accurate to within a few seconds per month. Some have batteries that take over during power outages. The 60-Hertz types should be checked periodically against an accurate time source such as WWV. Newly available are the inexpensive atomic clocks which automatically sync to WWV. Some may be affected when you transmit, so try out one before purchasing if you can borrow one. An indoor/outdoor thermometer is handy as (for what ever reason) Hams love to exchange weather reports and temperatures in C and F.

Desk references such as prefixes, Q-Signals, CW abbreviations, your data such as grid square, 10/10 number, longitude, latitude, etc. Also when you get a fair country count worked, a list of all DXCC countries and the ones you have worked by band and mode. That way you can quickly check if you need a particular country on a new band or mode. A QSL outstanding list will help you keep track of outstanding QSL's so that you can reapply before the logs expire.

A time zone clock is handy so you can let the DX know that you are aware of their time. A timer to alert you to ID. Most Hams over do the ID'ing, afraid of going past 10 minutes, so a timer may cut down on the constant callsign exchanges during an extended QSO.

NO NO and questionable gimmicks. On occasion you will hear microphones with a roger beep to signal an over to the other station. Drives some Hams batty – but you be the judge. Reverb units are a definite no no as is the mic gain wide open and talking across the room, hard to copy these.

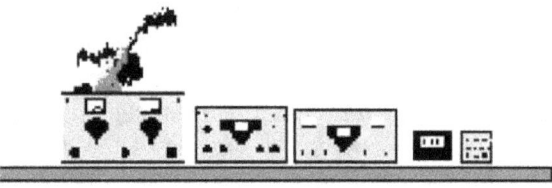

BUT I AM QRO

FAMOUS QUOTES FROM THE MOVIE - CASABLANKER	ROUND UP THE USUAL JAMMERS CLAUDE RAINS
	PLAY IT AGAIN SAM QSM OM

NOTES

CHAPTER III   OPERATING AIDS

3-1. CHAPTER CONTENTS

This chapter contains information on operating aids for DXing. The table below lists the chapter topics and paragraph numbers where they can be found.

PARAGRAPH & TOPIC	PARAGRAPH & TOPIC
3-2. Logging Programs	3-12. Macintosh Programs
3-3. Grid Squares	3-13. Propagation Websites
3-4. Distance & Bearing Programs	3-14. Propagation Software
3-5. Grayline Programs	3-15. Radio Control Programs
3-6. Antenna Programs	3-16. RTTY, PSK31 & SSTV Programs
3-7. NCDXF Beacon Programs	3-17. Prefix List
3-8. DXer Programs	3-18. CQ and ITU Zones
3-9. CW programs	3-19. Q-Signals & RST
3-10. Great Circle Programs	3-20. DXer Phonetics
3-11. Prefix Finder Programs	3-21. CW Abbreviations
	3-22. Other DX Books

3-2. LOGGING PROGRAMS

With the advent of the computer in the shack, it didn't take long for programs to appear to take the place of hand logging. Soon stations were reporting that they had worked before (Date, Time, Freq, QSL received, etc.) with a quick lookup in the computer. Some loggers are general purpose; others are whizbang for contesters, so try the free evaluation downloads that are available for most of these programs. Some of loggers are free, others are shareware, and of course some are buy ware. In addition to the 50+ logging programs found at QRZ.COM, others are at URL: http://ac6v.com/logging.htm

AALog	M*LOG	WINEQF
AGW Software	R25Log Logging Program	WJ2O logger
County Hunting Logging	ShackLog	
Dxbase	LOGPlus	**For DXers and**
Dxtreme	Log2000	**Contesters**
EUROWINLOG	Log Windows	AALog
Field Day Logger	Logger	AR Technology
HamLog for Macintosh	PROLOG	BV6QSL Label
Ham Radio Logbook	QQSL	GenerationCATCC
Hyper Log	Smart Log	CT by K1EA
LogEQF	Super WinLog	CQ Callbook
LOGic	SweepStakes Logger	DX Atlas
Lux-Log	TRX-Manager	DX Lab

DXLog and ClusterLog	LOGPlus	TR LOG
Dxtreme	LogWindows	Windows-QSL-Manager
HAMshack	NA Contest Logging	WriteLog
HamRadio Logbook	QRATE	Yplo
K1TTT Open Logger	RAC 'n Rollup	
K4HAV 10-10 Contest	SD Super Duper	

ALL of these can be found on the Internet at URL: http://ac6v.com/logging.htm and http://ac6v.com/dxsoft.htm Also see DX TelNet at URL: http://ac6v.com/dxcluster.htm

3-3. GRID SQUARES

The Maidenhead grid-square system, formalized at a VHF meeting in Britain in 1980 and adopted world-wide by the International Amateur Radio Union in 1985, is almost universally used as a locator system by VHF, UHF and microwave operators. The Maidenhead system divides the world into 32,400 squares, each 2 degrees of longitude by 1 degree of latitude. There are larger "fields" of 100 locator squares each, and each square is divided into smaller "subsquares." For most purposes, knowing your 2 degree by 1 degree square is sufficient. The Maidenhead Grid Square locator system has been long used by the VHFers, but more and more HFers are asking and chasing these. Another compelling reason for using grid squares is because many awards (ARRL and those from six meter organizations) are based on the number of grid squares worked.

For grid squares, locators and maps (USA), See URL: http://ac6v.com/opaids.htm#GRID

A neat program for locating grid squares is DX Atlas URL: http://www.dxatlas.com/ By VE3NEA. With a map displayed of the area desired, do a CNTRL G, then type in the grid square, a red 4-point locator will appear on the map so you can determine the state and location within the state. With this program, the state outlines are very clear.

Where the grid square is close to state borders, using the zoom function will enlarge the grid square to help determine a more precise location. In the case where the grid encompasses 2 or more states, you can use the RAC Flying Horse, Buckmaster or QRZ callbooks to determine the caller's state and city. (I have the RAC at the ready in my computer along with DX Atlas and a Logging program). Other useful tools to coordinate QSO's are the 50 MHz prop logger (http://dxworld.com/50prop.html) as well as the chat area at www.dxer.infc. Sometimes you only hear the callsign and want to know the city, state and grid square. Use the callbooks for this and hope the guy/gal didn't move.

One of the best grid square maps is from ICOM -- Click on AMATEUR, then U.S. Grid Square Map (470 KB PDF)

Artsci has a publication that although it is a repeater guide (excellent), it has a map of all USA states and the grid squares within that state. The map extends slightly beyond the state borders, so one can see a grid square that encompasses 2 or more states.

3-4. DISTANCE AND BEARING

In addition to your beam heading to a DXCC country, it's nice to know the distance between you and the DX. It makes a nice brag tape that you worked Bhutan with 100 milliwatts knowing that it is 10,000 km away at 295 degrees true!! Several programs or web sites are available to determine distance and bearing. These include:

Amateur Radio ADF
DXCC Entity List / Beam Headings, by Randy, W5UE
Distance and Bearing Finder, by Bali Online
Distance Finder, From Cid
Beam Headings to DXCC Countries From The NJDXA
WinHdg By WA6FHI
DX View 1.1.5 – Free Ware
GcmWin -- Via ftp sites
QTH Locator By Klaus Evers
Geodetic II -- Distance/Bearing Finder - From Klaus-Dieter Scholz
BeamHead.Zip

These programs can be found at URL: http://ac6v.com/opaids.htm

3-5. GRAYLINE

In general, grayline propagation occurs on the 160, 80, 40 and 30-meter bands. See Chapter 4, Propagation for details of this type of propagation.

Gray Line Program -- From PA3CQR   Gray Line Map – from spiritweb.org/
Gray Line Program -- From Alex Shovkoplyas   DX View 1.1.5 – from QSL.net
Gray Line Map -- from DX.QSL.NET   Gray Line Map -- by John Walker (Fourmilab)

All available at URL: http://ac6v.com/opaids.htm#GRAY

3-6. ANTENNAS

A whole raft of antenna programs is available from the Internet. These include;
   Antenna Design Freeware -- 1D and 2D arrays from MEI Software
   Antenna Shareware A Bunch
   Antenna Software A Ton -- From The Spread Spectrum Scene
   Antenna Software FreeBees -- Via Walter Banzhaf
   Antenna Links page -- From Andy Ketner - N3OGT
   Antenna Software by W7EL -- EZNEC, EZNEC Pro, ELNEC
   Antenna Software From QRZ
   Coax Loss Calculator -- From M5AHJ
   Cubical Quad Antenna Calculator -- From KD6DKS
   EZNEC 3.0 -- For Windows 95/98/NT/2000

Funet Antenna FTP Site
HamCalc By VE3ERP
Mac Antenna Master -- From Black Cat Systems
The MININEC Professional Series
NEC4WIN® -- Antenna Simulation Software
PolarPlot -- Polar diagram of your rotatable beam antenna from G4HFQ
Antenna programs for advanced quads, dipoles and more. VE3SQB
Yagi Designer -- Produces antenna radiation patterns and graphs
YagiMax 3.11
YagiStress -- By Kurt Andress, NI6W
YTAD -- by N6BV -- Determine antenna gain based on the slope of terrain
Quick Yagi -- Via WA7RAI's Home Page
 All available at URL: http://ac6v.com/antsoftware.htm

3-7. BEACONS

HF Beacons. There are hundreds of beacons on the air on almost all Amateur Bands. For an excellent listing on the internet, see BEACONS at URL: http://www.ac6v.com/

NCDXF International Beacon Network. The Northern California DX Foundation, in cooperation with the IARU, constructed and operates a worldwide network of high-frequency radio beacons on 14.100, 18.110, 21.150, 24.930, and 28.200 MegaHertz. These beacons help both amateur and commercial high-frequency radio users assess the current condition of the ionosphere. The entire system is designed, built and operated by volunteers at no cost except for the actual price of hardware components, shipping costs, and so on.

For those using the NCDXF Beacons, several free programs are available for displaying the beacon activity. These help beacon listeners figure out which beacon is transmitting on which frequency at any given time. The programs are great for people who can't easily copy Morse code at twenty-two words per minute.

They help determine the callsign of the very weak beacons so that you no longer have to try to simultaneously look at both your clock and your printed beacon schedule. It is imperative to have your computer clock synched to an accurate source (WWV, Dimension A, etc.) Beacon programs are available for a number of computers and operating systems as found on the Internet at URL http://ac6v.com/beacons.htm   The following table gives the minute and second within each hour of the start of the first transmission of each of beacon on each frequency. Each transmission is repeated every three minutes. A transmission consists of the callsign of the beacon sent at 22 words per minute followed by four one-second dashes. The callsign and the first dash are sent at 100 watts. The remaining dashes are sent at 10 watts, 1 watt and 0.1 watts.

NCDXF/IARU International Beacon Network

Slot	Location	Call	14.100	18.110	21.150	24.930	28.200	Operator	Status
1	UN	4U1UN	00:00	00:10	00:20	00:30	00:40	UNRC	OK
2	VE	VE8AT	00:10	00:20	00:30	00:40	00:50	RAC	OK
3	USA	W6WX	00:20	00:30	00:40	00:50	01:00	NCDXF	OK
4	KH6	KH6WO	00:30	00:40	00:50	01:00	01:10	NOARG, HARC	
5	ZL	ZL6B	00:40	00:50	01:00	01:10	01:20	NZART	-
6	VK	VK6RBP	00:50	01:00	01:10	01:20	01:30	WIA	OK
7	JA	JA2IGY	01:00	01:10	01:20	01:30	01:40	JARL	-
8	UA	RR9O	01:10	01:20	01:30	01:40	01:50	SRR	OK
9	VR2	VR2HK	01:20	01:30	01:40	01:50	02:00	HARTS	OK
10	4S7	4S7B	01:30	01:40	01:50	02:00	02:10	RSSL	OK
11	ZS	ZS6DN	01:40	01:50	02:00	02:10	02:20	ZS6DN	OK
12	5Z	5Z4B	01:50	02:00	02:10	02:20	02:30	RSK	OK
13	4X	4X6TU	02:00	02:10	02:20	02:30	02:40	U Tel Aviv	-
14	OH	OH2B	02:10	02:20	02:30	02:40	02:50	U Helsinki	OK
15	CS	CS3B	02:20	02:30	02:40	02:50	00:00	ARRM	-
16	LU	LU4AA	02:30	02:40	02:50	00:00	00:10	RCA	OK
17	OA	OA4B	02:40	02:50	00:00	00:10	00:20	RCP	OK
18	YV	YV5B	02:50	00:00	00:10	00:20	00:30	RCV	OK

3-8. DXER PROGRAMS

Besides the software for logging programs, there are several other programs, these include:
CT By K1EA
CATCC by N2CKH -- Computer Aided Transceiver Control Center
CLX -- Clone of AK1A Packet Cluster that runs under Linux
CLUSSE -- - DX cluster node software
DX Monitor -- DX announcements from OH2BUA
DX Packet Cluster Webnet -- AK1A, the Packet Cluster™ software package
Hamelot -- BEAM - displays world map

HAMshack v1.0 -- Full featured program, Rig Control, Logging, Contesting, Prefix list, Packet cluster, QSL printing, Awards tracking.

LogWindows -- From SCO -- combines logging, rig control, antenna rotor control, DX Packet cluster monitoring, and award tracking

Pile Up Tapes -- Via Contesting.com

G4ZFE Pile Up! Program

QQSL -- The Ultimate Ham Radio QSL Label Program by Bill Mullin - AA4M

RXCLUS -- free program for DX-cluster users -- From HB9BZA. A connect or no connect monitoring program.

UT1YV -- Software programs including: CW program, DSP filter program, radio logbook, Spectrum Analyzer, Beacon-Echo Repeater, for Windows 95/98/NT

TR Log

WinHdg -- By WA6FHI -- Find heading and country info from callsign

XPWare Communications Software By KF7XP -- Packet Programs, Transceiver control, DX Cluster monitoring

YPlog -- Radio Control and Logging Shareware de Tony, VE6YP

And much more at URL: http://ac6v.com/dxsoft.htm

3-9. CW PROGRAMS

In addition to the multitude of Morse training programs, there are programs for contesters and DXers:

PED - by MasIII H.Kozu - JE3MAS is one of the most popular band simulators. The new version simulates not only pile-up with up to 32 simultaneously calling stations but also noise, RTTY QRM and inaccuracy of sending at some stations. Another nice feature is paddle-keying facility. Requires a sound card.

Pile Up! - Windows pileup simulator by Richard Everitt, G4ZFE offers several modes of operation. There is a competitive list of stations having best scores. It has a wide range of selectable (adjustable) parameters.

HSTT - another powerful training program from the house of JE3MAS. Many modes, many features including Letter (characters only), Number (numerals only), Mix (a mixture of characters, numerals and symbols (? / , . =).

Rufz (random transmission of characters, numerals and slashes) (characters that appear in RUFZ) and Rufz (real) - call signs of actual stations taken from the RUFZFILEDTA of the RUFZ. Fully configurable with substitute keys, High Speed Morse (mastering tone color Morse) and much, much more. Developed for the JA team attending the HST Championships in HA and LZ, now a very sophisticated training system for all purposes.

RUFZ - another pile-up simulator by Mathias Kolpe, DL4MM is a powerful tool for those, who want to increase copying ability at very high speeds. One station is calling at a time but the program increases speed depending on number of errors made.

There is a large competitive list of stations having best scores. If you need to know what is your speed limit, try RUFZ. This program runs up to over 125 Wpm!

3-10. GREAT CIRCLE MAPPING

For those interested in Great Circle mapping, the Internet has several neat sites and programs. See URL: http://ac6v.com/software.htm#GEO

Amateur Radio ADF
Geo Clock World Maps, Gray Line -- much more By Joe Ahlgren
Gray Line Program -- From PA3CQR
Gray Line Program -- From Alex Shovkoplyas
DX Atlas
DX View 1.1.5 Find heading, country, distance, zone info from callsign prefix
Ham Clock Version 2, shows time, UTC and by prefix. Also displays the distance and heading from your location to the selected prefix next to the time.
Formulae For Calculating Great Circle Heading and Distance Information
Super Software Page - By KL7J, Great Circle Map, Geoclock w/Grayline, SSTV, RTTY
Great Circle calculations -- gcgc9900.zip - by Ron McConnell W2IOL.
Great Circle Map Generator Software. - From Brian Smith
Great Circle Maps - azprj104.zip program - For any point on the earth.
PCSW22 -- Distance/Bearing Finder -- - Download pcsw22.zip
Great Circle Distance and Bearing Calculator --On a Lotus spreadsheet -- - By David Stua
Online Map Creation using AZ_PROJ v1.04 - By Joe NA3T and Michael NV3Z
AZMap - an Azimuthal-Equidistance Map Generator From AA6Z
GcmWin - A program to make Great Circle Maps -- Via SM3GS
HamClock -- Desk Top program for Time in the All DXCC Prefixes
Dimension 4 -- Sets your PC Clock to world standards
Beacon Wizard -- Track the NCDXF Beacons and Sets your PC Clock to world standards
OSSYSTEMS -- Sets your PC Clock to world standards
Sunrise/Sunset, Gray line and UTC Time From HAB Software, Hamburg, Germany.

3-11. PREFIX FINDERS

Prefix finders can be found at URL: http://ac6v.com/zones5.htm#PR

AC6V's Mega Prefix List -- Prefixes By The Bushel
Great Prefix Finder -- From K4UTE
DXMonitor http://www.benlo.com/dxmon.html
Prefix Finder -- From HFRADIO.ORG
Prefix Lookup 1.0 Beta -- Download From Alex Shovkoplyas
Prefix Maps -- From VK2CA -- A must SEE
International Call-Sign Prefix Allocations -- From The ARRL
Cross Prefix List -- From The ARRL -- On the right in the Yellow Frame
Prefix List -- With Maps -- From The Elmer Hamlet
Table of Allocation of International Call Sign Series -- From The ITU
Another Mega Prefix List -- From The QSL Collection
Canadian Prefixes -- Special Prefixes and Call Signs
Russian Prefix/Oblast/Zone/Territory Lists
Territories, Prefixes, Suffixes and Zones of the Former USSR
Also see Appendix A10, this has a large collection of common, unusual, and changed prefixes.

3-12. MACINTOSH

Despite the PC users belief – MAC is alive and well in the Ham Radio World.
These are available at URL: http://ac6v.com/software.htm#MAC

3-13. PROPAGATION

Propagation websites can be found at URL: http://ac6v.com/propagation.htm
Websites include:

Propagation Mail List Reflectors -- Several to choose from
Basics of Radio Wave Propagation
Propagation Primer -- Excellent via HFRadio.org and AE4RV
Latest Solar Report -- Nice Charts and Graphs
Solar Activity -- Excellent From Dan, W3DF
Propagation Center -- Excellent From NW7US, Tomas
QST HF Propagation Prediction Charts -- MUF, Best Propagation Paths and Times -- PDF Files
-- From the ARRL
27 Day HF Propagation Prediction Chart -- From Paul, AA1LL
PROPAGATION -- From QSL.NET
Propagation Studies Committee - RSGB -- By Martin Harrison, G3USF
IPS Support for HF Radio in North America -- From Australia
IPS Support for HF Radio World Wide -- From Australia
Non-Ducting Tropo Propagation -- From M1BWR

Radio Propagation Sources -- From Radio Netherlands
All About Sunspots & The Solar Cycle -- From sunspot.com
Predicted Solar Flux and Planetary A Index over 30 days. -- From NWRA Space Weather Links
Skywave Technologies -- Propagation Tutorial & Programs
Solar Activity -- Via NB6Z
Radio Propagation Conditions From Ham Radio On-Line
Propagation Topics -- Via The Space Environment Center -- Includes a GLOSSARY OF SOLAR-TERRESTRIAL TERMS
Propagation Glossary -- From The Basics of Radio Wave Propagation Pages
ARRL Propagation Forecasts.
Solar flux A & K -- Summaries by calendar quarter --1994 to 1997
Solar flux A & K -- Current Months Solar data
Solar Flux Graphs --By WN6K, Paul
Solar Flux Graphs -- from The CT1BOH Web Site
Solar Images -- Fantastic Shots of Ol Sol
Solar Terrestrial Activity Report -- By Jan Alvestad
Long Delayed Echo (LDE) -- From EA6VQ

3-14. THE PROPAGATIONSOFTWARE PAGES

Solar Data Plotting UTILITY -- By Scott Craig
Kangaroo Tabor Software -- WinCAP Wizard, CAPMan, Active Beacon Wizard -- Propagation and Beacon
VOACAP -- Voice of America - Propagation Prediction Program
High Frequency Propagation Models -- U.S. Department of Commerce
PP - Propagation Prediction Version 1.5 -- From Bavarian Contest Club
KU5S -- Communications Analysis Prediction Wizard
Skywave Technologies -- Propagation Tutorial & Programs
Gray Line Map Program -- By PA3CQR
HF Prediction Software -- From IPS Radio & Space Services
K9SE Propagation Assessment Software Page
Visual Meteor --Download latest V4.3 release 3 VM-Soft
VHF Solar Activity, Aurora & Meteor Scatter Pages
The Aurora Page
High Frequency Active Auroral Research Program (HAARP)
QSLing HAARP

All available at URL: http://ac6v.com/propagation.htm

3-15. RADIO CONTROL

Modern rigs can be controlled by computer which offers very powerful and versatile features. Some samples of what is available:

AMREG -- FREE downloadable AMREG software for ICOM radios
AOR Radios Control Software -- AR5000 AR8000 AR8200
CAT link software. -- For the YAESU FT-736R and FT-847
CI-V Commander Control for your Icom radio
Control Commander -- For the Kenwood TS-570 & Others
DXtra Inc. -- For The TenTec RX-320
FineWare -- Control for a variety of SWL Receivers
KingSmith Software -- Yaesu FT-736R, and Yaesu FT-847, ICOM IC-746, ICOM IC-756, and Yaesu FT-920.
MacIcomControl -- Communicate with and control an Icom radio with a McIntosh
MacSchedKeeper -- Schedule your Kenwood TS-50 for SWL Schedules or Ham Nets with a MacIntosh
Micro-G2b Rotor control -- Control up to 4 rotors
Radio Master 2000 -- Kenwood Radio Control program.
Scan920 -- a radio control/scanner program for Yaesu FT-920 From KH2D
TRX-Manager -- Yaesu, Icom, Kenwood and TenTec Radios
Radio Manager -- For Scanners
Rig Control Programs From QRZ.com
All available at URL: http://ac6v.com/software.htm#RC

3-16. RTTY, PSK31, AND SSTV PROGRAMS

RTTY -- teletype contesters program by WF1B
RTTY Loop Software -- From Marc I. Leavey, M.D. WA3AJR
Many, many PSK31, MMTTY, MFSK16, are all gaining in popularity
TrueTTY 0.10 -- From UA9OSV -- receive RTTY via a sound card.
N3TUK Software Pages -- RTTY, CW, Fax & SSTV
SSTV for Windows 95/NT Public Beta version 1.04
Slow Scan TV Programs From QRZ.com
Mscan v3.11 -- SSTV program for Windows 95/98 offers sound card support
Mscan Meteo Text -- Windows program to receive NAVTEX and RTTY.
See URL: http://ac6v.com/software.htm

3-17. PREFIX LIST

Most often a prefix can be found in the ITU lists in the Appendices. For many years now, the author has tracked the myriad of prefixes that one hears on the air. These are presented in Appendix A10.

3-18. CQ AND ITU ZONES

There are two major zone systems in effect for DXing, The CQ Zones and the ITU Zones. There are 40 CQ Zones and about 75 land-based ITU zones. Maps are shown in Appendices A6 and A7.

3-19. Q-SIGNALS AND RST REPORTING

Q-Signals are in Appendix A5 and the RST Reporting System is in Appendix A3.

3-20. DXer PHONETICS

For AC6V's unofficial list of DXer phonetics, see Appendix A4.

3-21. CW ABBREVIATIONS

These can be found in Appendix A2

3-22. OTHER DX BOOKS

Low Band DXing By ON4UN URL: http://www.universal-radio.com/catalog/books/3635.html

The Complete DX'er By W9KNI URL: http://www.idiompress.com/books-complete-dxer.html

NOTES

CHAPTER IV   PROPAGATION

4-1. CHAPTER CONTENTS

This chapter contains information on propagation for DXing. The table below lists the chapter topics and paragraph numbers where they can be found.

PARAGRAPH & TOPIC	PARAGRAPH & TOPIC
4-2. The Regions Of The Ionosphere	4-13. 80 Meters Propagation
4-3. Ionosphere Variations	4-14. 60 Meters Propagation
4-4. Solar Flux, A Index, K Index	4-15. 40 Meters Propagation
4-5. Gray Line DXing	4-16. 30 Meters Propagation
4-6. Backscatter	4-17. 20 Meters Propagation
4-7. Sporadic E	4-18. 17 Meters Propagation
4-8. Tropospheric Scatter	4-19. 15 Meters Propagation
4-9. Ducting	4-20. 12 Meters Propagation
4-10. Beacons	4-21. 10 Meters Propagation
4-11. Short Wave Stations	4-22. Propagation Graphs
4-12. 160 Meters Propagation	4-23. Six Meter Propagation

The key to DXing in order is probably:
1. Propagation
2. Propagation and more Propagation
3. Good DX Antenna (low angle radiator for the upper HF bands)
4. Operating Skills (stealth will overcome brawn as in many disciplines)
5. Transceiver (A Swan 350 will do) but nice to have better selectivity, etc.
6. Linear (two S-Units increase from 100W to 1500W)

4-2. THE REGIONS OF THE IONOSPHERE

Propagation is a highly complex subject and many volumes of radio propagation information have been written over the years. Three excellent texts are (1.) The Radio Amateurs Guide to the Ionosphere by Leo F. McNamara (formerly with the Australian Ionospheric Prediction Service), Krieger Publishing Company; ISBN: 0894648047 (2) The Little Pistol's Guide to HF Propagation by Bob Brown NM7M (retired physics professor whose specialty was ionospheric physics), (3) The New Shortwave Propagation Handbook by Jacobs W3ASK, Cohen N4XX, and Rose K6GKU available from CQ Communications. These are detailed, but not lots of math. Following is a practical layman's view of propagation. It is intended for the average DXer, but the reader is encouraged to read the texts mentioned above for a more thorough understanding of the mystical world of propagation.

The ionosphere is a region extending from a height of about 30 miles to over 310 miles.

Here some of the molecules of the atmosphere are ionized by radiation from the sun to produce an ionized gas. Electrons are produced when radiation collides with uncharged atoms and molecules. Since this process requires solar radiation, production of electrons mostly occurs in the daylight hemisphere of the ionosphere. The exception is solar radiation scattered into the dark ionosphere by the Earth's geocorona, resulting in residual nighttime E region ionization.

Ionization is the process in which electrons, which are negatively charged, are removed from (or attached to) neutral atoms or molecules to form positively (or negatively) charged ions and free electrons. It is the ions that give their name to the ionosphere, but it is the much lighter and more freely moving electrons which are important in terms of high frequency (3 to 30 MHz) radio propagation. Generally, the greater the number of electrons, the higher the frequencies that can be used.

During the day there may be four regions present, the D, E, F1 and F2 regions. Their approximate height ranges are:

> D region 30 to 55 miles
> E region 55 to 87 miles
> F1 region 87 to 130 miles
> F2 region over 130 miles

During the daytime, sporadic E propagation is sometimes observed in the E region, and at certain times during the solar cycle the F1 region may not be distinct from the F2 region but merge to form an F region. During the daylight hours, the D region absorbs radio energy below 10 MHz depending on density. At night the D region essentially disappears and the E region reaches its lowest ionization, leaving only the F2 region available for reliable long distance communications, but sporadic E may occur into the night. Only the E, F1, F2, and sporadic E refract HF waves.

The F2 region is the most important for high frequency radio propagation as it is present 24 hours of the day and its high altitude allows for the longest communication paths. The F2 layer usually refracts the highest frequencies in the HF range. The lifetime of electrons is greatest in the F2 region which is one reason why it is present at night. Typical lifetimes of electrons in the E, F1 and F2 regions are 20 seconds, 1 minute and 20 minutes, respectively.

4-3.   IONOSPHERE VARIATIONS

The ionosphere is not a stable medium which allows the use of a frequency over the course of a year, or even over 24 hours. It varies with the solar cycle, the seasons, the paths, and can vary from day to day. A frequency that may provide successful propagation may not do so an hour later.

The sun exhibits a long-term periodic rise and fall in activity which has a profound effect on HF communications. **This is termed the solar cycle which historically has varied in time from 9 to 14 years, 11 years being typical.**

27-DAY SUNSPOT CYCLE.—The number of sunspots in existence at any one time is continually subject to change as some disappear and new ones emerge. As the sun rotates on its own axis, these sunspots are visible at 27-day intervals, the approximate period required for the sun to make one complete rotation. The 27-day sunspot cycle causes variations in the ionization density of the layers on a day-to-day basis. The fluctuations in the F2 layer are greater than for any other layer. For this reason, precise predictions on a day-to-day basis of the critical frequency of the F2 layer are not possible. In calculating frequencies for long-distance communications, allowances for the fluctuations of the F2 layer must be made.

At the 11 year solar minimum, only the lower frequencies of the HF band will be propagated by the ionosphere, while at solar maximum the higher frequencies are supported. At solar maximum, there is increased radiation from the sun, producing more electrons in the ionosphere which allows the use of higher frequencies. Summertime propagation is generally not as good for HF communications as spring or fall. In summer, high noise levels and shorter periods of darkness in the northern hemisphere make propagation on 160 and 80 meters difficult.

When the sun is highly active, there is a greater likelihood of large solar flares occurring. Flares are huge eruptions on the sun which emit radiation at X-ray wavelengths that can ionize the D region causing increased absorption of HF waves. Since the D region is present only during the day, only those communication paths, which pass through daylight, will be affected. The absorption of HF waves traveling via the ionosphere after a flare has occurred is called a short wave fade-out. Fade-outs occur instantaneously and have a greater affect on the lower frequencies. Lower frequencies are also the last to recover. If a flare is very large, the whole of the HF spectrum may be rendered unusable. The duration of fade-outs can vary between about 10 minutes to over an hour depending on the intensity and duration of the flare.

E region frequencies are greater in summer than in winter, however, the variation in F region frequencies is more complicated. In both hemispheres, F region noon frequencies generally peak around the March and September equinoxes. Around solar minimum, the summer noon frequencies are generally greater than those in winter, but around solar maximum, winter frequencies at certain locations, can be higher than those in summer. In addition, frequencies around the equinoxes are higher than those in summer or winter for both solar maximum and minimum. The observation of noon winter frequencies often being greater than those in summer is called the seasonal anomaly.

There are variations in the E and F region frequencies at noon and midnight from the poles to the geomagnetic equator. During the day and with increasing latitude, solar radiation strikes the atmosphere more obliquely, so the intensity of radiation and the electron density production decreases towards the poles. The effects of latitude can have a significant bearing on reception. Amateurs in the Northwest USA, Canada, and Alaska are often jealous of operators at lower latitudes, who frequently get better openings on the higher frequencies. Northwest USA hams are known as the Suffering Sevens.

Operating frequencies are normally higher during the day and lower at night. With dawn, solar radiation causes electrons to be produced in the ionosphere and frequencies increase reaching their maximum around noon.

During the afternoon, frequencies begin decreasing due to electron loss and with evening, the D, E and F1 regions become insignificant. HF sky wave communication during the night is therefore by the F2 region and absorption of radio waves is lower because of the lack of the D region. Through the night, frequencies decrease reaching their minimum just before dawn. The D region attenuates waves as they pass through it, but becomes insignificant at night. Absorption on the lower frequencies at night is in the lower E region.

Absorption also varies with the solar cycle, being greatest around solar maximum. Signal absorption is greater in summer and during the middle of the day. There is a variation in absorption with latitude, with more absorption occurring near the equator and decreasing towards the poles, although certain solar events will significantly increase absorption at the poles. This accounts for the so-called day and night time bands.

Around the polar regions, absorption can affect communications quite drastically at times. High-energy protons ejected from the sun during large solar flares will migrate down the Earth's magnetic field lines and into the Polar Regions. These protons can cause increased absorption of HF radio waves as they pass through the D region. This increased absorption may last for a number of days and is called a Polar Cap Absorption event (PCA).

Spread F occurs when the F region becomes diffused due to irregularities in the region, which scatters the radio waves. The received signal is the superposition of a number of waves refracted from different heights and locations in the ionosphere at slightly different times. At low latitudes, spread F occurs mostly during the night hours and around the equinoxes. At mid-latitudes, spread F is less likely to occur than at low and high latitudes and more likely to occur at night and in winter. At latitudes greater than about 40 degrees, spread F tends to be a night time phenomenon, appearing mostly around the equinoxes, while around the magnetic poles, spread F is often observed both day and night. At all latitudes there is a tendency for spread F to occur when there is a decrease in F region frequencies. Spread F is often associated with ionospheric storms.

High frequency radio signals (3 to 30 MHz) can propagate via ground wave, line-of-sight, or by sky wave propagation. Ground wave travels short distances, about 60 miles over land and 180 miles over sea. The range of the ground wave depends on antenna height, polarization, frequency, ground types, vegetation, terrain and sea conditions.

Direct or line-of-sight waves may interact with the earth-reflected wave depending on terminal separation, frequency and polarization.

Sky waves are refracted by the ionosphere and can achieve extremely long distances. However, not all HF waves are refracted by the ionosphere; there are upper and lower frequency limits for communications between two points. If the frequency is too high, the wave will penetrate the ionosphere, if it is too low, the strength of the signal will be lowered due to absorption in the D region. The range of usable frequencies will vary throughout the day, the seasons, the solar cycle, and from location to location depending on the ionospheric region that is effective for communications. While the upper limit of frequencies varies mostly with these factors, the lower limit is also dependent on receiver site noise, antenna efficiency, transmitter power, E layer density and absorption by the ionosphere.

For any path there is a Maximum Usable Frequency (MUF) that is determined by the state of the ionosphere in the vicinity of the refraction areas and the length of the path. The MUF is refracted from the area of maximum electron density of a region. Therefore, frequencies higher than the MUF for a particular region will penetrate the region. During the day it is possible to communicate via both the E and F layers using different frequencies.

The highest frequency supported by the E layer is the EMUF, while that supported by the F layer is the FMUF. The F region MUF varies during the day, seasonally and with the solar cycle.

When a refracted radio wave returns to earth it reflects upwards again and results in the wave making multiple hops. The hop length is the ground distance covered by a radio signal after it has been refracted once from the ionosphere and returned to Earth.

The upper limit of the hop length is set by the height of the ionosphere and the curvature of the Earth. For E and F region heights of 60 miles and 186 miles, the maximum hop lengths with an elevation angle of 4 degrees, are 1000 miles and 2000 miles, respectively. Distances greater than these require more than one hop. For example, a distance of 4000 miles would require a minimum of 4 hops by the E region and 2 hops via the F region with the foregoing elevation angle. And of course, the propagation mode with more hops may be required using antennas that don't put out much energy at low elevation angles.

Aurora propagation occurs when more than the usual levels of charged particles arrive at earth forming an auroral E layer which may reflect radio waves from the HF-band (3-30MHz) all the way up to and including the entire UHF-band (300-3000MHz). Heavy fading (QSB) is common with these signals. An auroral signal is easily recognized at 30MHz as a bubbling sounding modulation or "under-water-like" modulation. Finally, because of the extreme and sudden phase shifts, narrow band modes such as CW and digital are the most reliable modes for DX contacts.

Trans-Equatorial propagation occurs during the spring and fall months, being best at higher sunspot numbers. This form of propagation allows two stations at nearly identical middle latitudes on opposite sides of the geomagnetic equator to communicate at frequencies up to 150 MHz. For example, communications can occur between Italy and South Africa or between the West Indies and South America.

Trans-Atlantic propagation is a rare type of propagation occurring between Europe and North America during the summer months, at a sunspot minimum, and well after sunset. These have occurred at times when DX contacts across the Atlantic seemed impossible. The mechanism of propagation is still unclear, but one proposed theory suggests that a gigantic Es-cloud forms above the entire Atlantic resulting in sky wave propagation.

Old timers can quote you chapter and verse about trans-equatorial paths, gray-line paths, seasonal paths, etc. But there are easier ways to determine propagation. Here we will look at propagation aids that will help you work DX. One of the most valuable sources of propagation information is the NCDXF/IARU International Beacon Network – see Beacons below.

Another useful source is the QST propagation charts but unfortunately they are no longer published in QST magazine, but are available on the ARRL Web Site at URL: http://www.arrl.org/qst/propcharts. Here we can see an optimum time at about 1800 UTC.

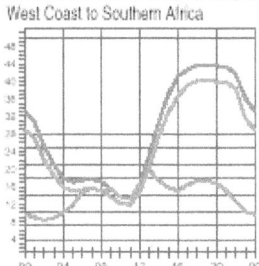

West Coast to Southern Africa

With these you can see propagation predictions from your QTH to other parts of the world. Included are GMT times and frequency. The monthly median MUF (Maximum Useable Frequency) curves are shown indicating the highest expected frequencies of propagation.

The LUF curves are useful for the lower frequency bands. For example in February 2001, the path from the West Coast USA to Southern Africa peaks at about 1700Z and the MUF is way up around 44 MHz. These paths change month by month as the earth travels around the sun and of course depends on the point of the solar cycle we are in. The daily SFI, A, K values can have a significant influence as well.

Another good source is the ARRL Propagation Bulletins, which as an ARRL member you can subscribe to and receive periodically from the league. In addition to the SFI, A, and K indices, a typical bulletin will include predicted conditions from the USA to the rest of the world. Example:

To Western Europe
80 meters 2230-0800z (best 0230-0630z)
40 meters 2130-1100z (best 0030-0730z)
30 meters 2030-1200z (best 2330-0730z)
20 meters 1400-0030z
17 meters 1330-0030z
15 meters 1430-2030z
12 meters 1500-1930z
10 meters 1600-1830z

To get deeper involved in propagation predictions, the IONCAP program will yield much information. This is "The Ionospheric Communications Analysis and Prediction Program" (IONCAP) and provides a means to calculate HF propagation parameters at any location. Field strength, mode reliability, the maximum usable frequency (MUF) and the lowest useful frequency (LUF) are some of the parameters calculated by IONCAP. They enable the user to specify antenna gains as a function of take-off angle and to specify required system performance in terms of the signal-to-noise ratio evaluated at the receiving point of the circuit. Required inputs include the transmitter and receiver location, transmitter power, universal time, month, and sunspot number.

A friendly version of IONCAP is WINCAP Wizard by Kangaroo Tabor Software. This reduces much of the jargon down to useable terms. Wizard generates reports based on your station power and antenna gains. The "Best Band Report" specifies the best ham band for a given hour. It can be found at URL: http://www.taborsoft.com/wwizard/

Another excellent program is W6ELProp™ at URL: http://www.qsl.net/w6elprop/ W6ELProp predicts ionospheric (sky-wave) propagation between any two locations on the earth on frequencies between 3 and 30 MHz. There is no charge for W6ELProp when used for non-commercial purposes.

All of these propagation prediction programs are based on a correlation between smoothed sunspot number (SSN – a 12-month running average) and monthly median ionospheric parameters. As such, the outputs of propagation prediction programs (usually MUF and signal strength) are statistical in nature – they are not absolutes. The reported values are median values, meaning that the reported MUF or signal strength should occur on at least half the days of the month. Since the model of the ionosphere in these programs is based on SSN, it or the equivalent smoothed 10.7cm solar flux should be used for best accuracy.
Using the daily 10.7cm solar flux reported by WWV compromises prediction accuracy, as the ionosphere really does not react on a one-to-one basis to the small daily variations of the sun.

4-4. SOLAR FLUX, A INDEX, K INDEX

Propagation is tied to the number of sunspots on the surface of the sun, since the areas around sunspots emit large amounts of ionizing radiation - extreme ultraviolet radiation. Increased sunspots correlate closely with better worldwide radio propagation. When there are more sunspots, the sun emits radiation that charges particles in the earth's ionosphere. Radio waves bounce off of (refract from) these charged particles, and the denser these clouds of ions, the better the HF propagation. The sunspot numbers are calculated by counting the spots on the visible solar surface and also by measuring their area.

Listening to WWV or checking propagation web sites will give the latest solar-terrestrial indices. It includes the 10.7cm solar flux index, Boulder A index and the Boulder K index. Solar Flux at 10.7cm is essentially a measurement of the thermal radiation of the sun, and contributes nothing to the ionization process. Solar flux is measured at several points on the earth, for example one is an observatory in Penticton, British Columbia using an antenna pointed toward the sun connected to a receiver tuned to 2.8 GHz, which is at a wavelength of 10.7 cm. The 12-month running average of 10.7cm solar flux correlates very well with the 12-month running average of the sunspot number – called the smoothed sunspot number and abbreviated SSN. The higher the smoothed SFI number, the better. Typical daily SFI values have ranged from 67 (Jan 1997) to 370 (Jan 1991).

Other solar activity of concern to HF operators are solar flares and coronal holes, which can emit energetic protons and X-rays and cause a significant increase in the solar wind speed. Energetic protons can cause polar cap absorption events (PCAs). X-rays can cause blackouts on the daylight side of the Earth due to increased absorption in the D region. And a significant increase in solar wind speed can result in geomagnetic storms that generally depress MUFs.

The A Index is an averaged quantitative measure of geomagnetic activity derived from a series of physical measurements.

The Boulder A index in WWV announcements is linear in nature and ranges from 0 to 400, and is the 24-hour A index derived from the eight 3-hour K indices recorded at Boulder, Colorado. The K index is logarithmic in nature and ranges from 0 to 9, and is the result of a 3-hourly magnetometer measurement comparing the current geomagnetic field orientation and intensity to what it would have been under geomagnetically quiet conditions.

Suffice it to say that the geomagnetic activity, solar storms, X-Rays, flares, etc., can have an adverse effect on propagation. The Planetary A index relates to geomagnetic stability. Magnetometers around the world are used to generate a number called the Planetary K index. A one-point change in the K index is quite significant. K index readings below 3 generally mean good stable conditions, and above 3 can mean high absorption of radio waves. Each point change reflects a significant change in conditions. Generally the higher the latitude of the measuring station, the higher the K and A indices reported. This is because the effects of geomagnetic instability tend to concentrate toward the polar regions of the globe.

Oversimplification can be very misleading in the complex field of propagation, but in general for long distance HF, the rule of thumb is the higher the SFI and the lower the A and K indices, the better the conditions on the higher frequencies. The A index should be under 14, and the solar activity low to moderate. If the A-index drops under 7 for a few days in a row and the SFI is up, watch for some really exciting intercontinental conditions.

You can hear the SFI, A and K indices on WWV or WWVH or by calling 303-497-3235. Geophysical alerts are broadcast from WWV at 18 minutes after the hour and from WWVH at 45 minutes after the hour. Both stations operate in the high frequency (HF) portion of the radio spectrum. WWV and WWVH radiate 10,000 W on 5, 10, and 15 MHz. The radiated power is lower on the other frequencies: WWV radiates 2500 W on 2.5 and 20 MHz while WWVH radiates 5000 W on 2.5 MHz and does not broadcast on 20 MHz. Each frequency is broadcast from a separate transmitter. Although each frequency carries the same information, multiple frequencies are used because the quality of HF reception depends on many factors such as location, time of year, time of day, the frequency being used, and atmospheric and ionospheric propagation conditions. The variety of frequencies makes it likely that at least one frequency will be usable at all times.

These values are also available on all DX Packet Clusters (SH/WWV). Also you can find them on numerous pages on the internet.

Note that the K index reported on WWV is only updated every 3 hours and the A and SFI values are usually updated once a day at 21 UTC. The A index reported at 21 UTC is calculated from the last eight K index readings in Boulder, so it contains data that is 24 hours old when it is first posted.

For graphs that update every 5 minutes see http://www.sec.noaa.gov/today.html these show short term events long before they show in the WWV report.

Also watch the K index graph at NOAA; it often differs from the WWV report, because the web page shows the estimated Kp (planetary K index) rather than just the reading from one site (Boulder) that is reported on WWV.

Classification of K-indices is as follows:

K0 = Inactive	K5 = Minor storm
K1 = Very quiet	K6 = Major storm
K2 = Quiet	K7 = Severe storm
K3 = Unsettled	K8 = Very severe storm
K4 = Active	K9 = Extremely severe storm

As with the K-index, the higher the A-index, the more unstable propagation becomes. Classification of A-indices is as follows:

A0 - A7 = quiet
A8 - A15 = unsettled
A16 - A29 = active
A30 - A49 = minor storm
A50 - A99 = major storm
A100 - A400 = severe storm

And another expert advises the summer seasonal reduction in Maximum Usable Frequency isn't caused by heating, but by a chemical difference in the F-layer during summer. You can check this with MINIPROP, taking a path from W7 to EU, summer and winter. This is covered in greater detail in "The Little Pistol's Guide to HF Propagation by NM7M.

4-5. GRAY LINE DXing

Gray Line is the area along the sunrise/sunset terminator that is in twilight. The terminator is the line around the world that divides day from night. Propagation under gray line conditions works well, especially for propagation to near the antipode (the location on Earth diametrically opposite to your QTH). An important characteristic of gray line propagation on the low bands is signal strength enhancement at the sunset end of the path (occurs occasionally) and at the sunrise end of the path (usually the best). Propagation along the grey line is very efficient. One major reason for this is that the D layer, which absorbs HF signals, disappears rapidly on the sunset side of the gray line, and it has not yet built upon the sunrise side.

To use the Gray Line, you will need sunrise/sunset information, these are on the Internet, see Chapter 3, DXing Aids. Also one can use the mapping feature of W6ELProp (free download at www.qsl.net/w6elprop), a propagation prediction program that produces a table of working frequencies along sunrise and sunset times, lengths of short path and long path, number of hops, angle of radiation, a frequency diagram, a compass rose of frequencies, and much more. Also the gray line can be seen at URL: http://dx.qsl.net/propagation/greyline.html

If you want to get into low band DXing using the gray line - see "Low Band DXing", by John Devoldere, ON4UN - highly recommended. Available through ARRL Bookstore, HRO, AES and most Ham stores.

4-6. BACKSCATTER

Backscatter propagation generally occurs when the maximum usable frequency (MUF) rises above 30MHz. When radio waves reach the ionosphere, they are reflected back towards the earth's surface at a larger detectable angle than usual. Backscatter signals reflect at a very sharp angle back toward the transmitting station and can cover the range of the usual ground wave and may be heard out to 1500 miles from the transmitting station. These are characterized with a distinctive hollow sound and exhibit very little fading, although they are usually much weaker than normal reflected waves. During periods of low solar flux, it requires a directional antenna to produce readable signals. During periods of very high solar flux, stations using low power and a vertical antenna may be successful using backscatter. In some instances, pointing the beam 180 degrees away from the desired station can achieve the desired contacts.

4-7. SPORADIC E

Sporadic E propagation takes place with irregular localized patches of relatively high electron density in the E region of the ionosphere.

The sporadic E is a regular daytime occurrence over the equatorial regions and is common in the temperate latitudes in late spring, early summer and, to a lesser degree, in early winter. Sporadic E can occur during daytime or nighttime, and it varies markedly with latitude. It can be associated with thunderstorms, meteor showers, solar activity and geomagnetic activity. This form of propagation usually occurs on the higher frequencies and single hops of 700 - 1500 miles can be achieved. Sporadic E openings on 6, 10, and 12 Meters are a great way to work the areas closer in to your QTH.

4-8. TROPOSPHERIC SCATTER

This is the only form of propagation that is directly influenced by the surface weather of the earth. The troposphere (0- 6 miles altitude) is composed of layers of air having different temperatures and moisture contents. When a sharp transition, called an inversion, appears between a cold dry layer and a warm moist layer of air, this transition causes refraction of radio waves.

This same type of refraction occurs when a radio wave travels through a climate inversion; if the inversion is strong enough, radio waves can be refracted back to earth after traveling significant distances. This propagation effect is seen most often in the VHF and UHF bands, especially the 6-meter band. (Up to several hundred miles).

4-9. DUCTING

On rare occasions, two or more inversions may appear at different altitudes and may allow some radio waves to be conducted between these two inversions. Records of over 1500 miles have been recorded as a result of ducting on VHF and UHF.

Usually confined to 2 meters, but it can occur as high as 1.2 GHz (usually along frontal systems), and it almost never occurs below frequencies of 50MHz.

4-10. BEACONS

Radio Beacons are the lighthouses of the airways. Usually unattended, they send out a radio signal periodically identifying their callsign. From these one can determine propagation conditions around the world. A list of radio beacons is available on the Internet. See URL: http://ac6v.com/beacons.htm But the premier beacon system for Amateur Radio is the NCDXF/IARU International Beacon Network. The Northern California DX Foundation, in cooperation with the International Amateur Radio Union, has constructed and operates a worldwide network of high-frequency radio beacons on 14.100, 18.110, 21.150, 24.930, and 28.200 MHz. Equipment used at each beacon site includes a Kenwood TS-50S transmitter, a Cushcraft R-5 vertical antenna, a Trimble Navigation Acutime (TM) GPS receiver (recently updated and renamed the Palasade (TM)) and a NCDXF controller.

Currently there are 18 beacons in operation around the world. Each beacon takes a turn in transmitting for a period of 10 seconds, so with 18 beacons, the system takes 3 minutes to complete a system cycle. For example at 14.100 MHz at the start of the hour, the United Nations beacon 4U1UN at the United Nations Headquarters in New York transmits for 10 seconds, then the Canadian beacon VE8AT follows for 10 seconds, followed by beacons in United States (CA), Hawaii, New Zealand, Australia, Japan, Russia, Hong Kong, Sri Lanka, South Africa, Kenya, Israel, Finland, Madeira, Argentina, Peru, and Venezuela. They transmit in the order listed. The beacons transmit every three minutes, day and night. A transmission consists of the callsign of the beacon sent at 22 words per minute followed by four one-second dashes. The callsign and the first dash are sent at 100 watts. The remaining dashes are sent at 10 watts, 1 watt and then 0.1 watts.

By listening for three minutes on a particular band, one can determine band openings to different parts of the world. By checking the other bands, a determination can be made as to which band has the best propagation to a particular part of the world. If your code speed is fast enough and the beacon can be heard, one can simply listen on the beacon frequencies and copy the callsigns. However, because the beacons transmit at known times, it is easy to determine which beacon that one is hearing without actually copying the CW callsign.

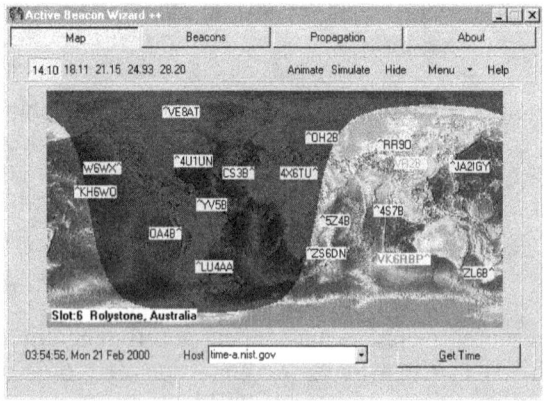

Since the beacons are running one hundred watts to a vertical, even a weak beacon signal may indicate a path with excellent propagation for stations using higher power and directive antennas. In addition, several computer programs are available that give a visual display of the beacon system, complete with the beacon callsigns and location. Some of these are automated such as BeaconSee. The schedule for the beacon system can be found on the Internet at URL: http://www.ncdxf.org/beacon/beaconSchedule.htm Also see Chapter 3.

4-11. SHORT WAVE STATIONS

Almost all modern Ham transceivers cover the entire HF band allowing for excellent listening on the short wave bands. In addition to beacons on the amateur radio bands, it is instructive to check on short wave stations which in general are in the following bands:

BAND		BAND	
120 m	2300 -- 2495 kHz	25 m	11.600 -- 12.100 MHz
90 m	3200 -- 3400 kHz	22 m	13.570 -- 13.870 MHz
75 m	3900 -- 4000 kHz	19 m	15.100 -- 15.800 MHz
60 m	4750 -- 5060 kHz	16 m	17.480 -- 17.900 MHz
49 m	5900 -- 6200 kHz	15 m	18.900 -- 19.020 MHz
41 m	7100 -- 7350 kHz	13 m	21.450 -- 21.850 MHz
31 m	9400 -- 9900 kHz	11 m	25.600 -- 26.100 MHz

For a listing of short wave stations broadcasting in English and their time schedules, see URL: http://ac6v.com/swll.htm includes almost all countries in the world from Albania to Zaire.

4-12. 160 METERS PROPAGATION

160 meters. Daytime conditions for this band suffer from extreme D-layer absorption, reducing the amount of signal to levels far below the noise floor of our receivers. This limits daytime coverage to essentially ground-wave coverage. At night, the D layer dissipates rapidly and worldwide 160-meter communication becomes possible via the F2-layer and in ducts in the electron density valley above the E region peak. Depending on the propagation mode, high or low elevation angles may be required. A limiting factor is the noise levels prevalent at these frequencies, both atmospheric and man-made as well as tropical and mid-latitude thunderstorms which cause high levels of static in the summer season. Winter conditions are much better, making winter evenings the best time to work 160-meter DX.

Special antennas schemes, such as loops or beverages, can significantly reduce the amount of received noise while enhancing desired signals. For DXing, power and amplifiers become a factor on the lower bands 160 to 40 meters, the 100-Watt transceivers not easily getting above the noise.

4-13. 80 METERS PROPAGATION

80 meters is similar to the 160-meter band. Here again, daytime absorption is significant and ground wave and some sky-wave propagation make communication ranges out to 250 miles possible. At night, signals can propagate halfway around the world.
Like 160 meters, 80-meter atmospheric noise is a problem, making the winter season the most attractive for the DXer. See 160 meters for other considerations. Local sunrise is a good time to work DX on the lower bands. Predictions are quite complex – but see ON4UN's book for a great treatment of low band Dxing.

4-14. 60 METERS PROPAGATION

The new 60 Meter band has characteristics similar to the 80 and 40 Meter bands.
Typical winter time results are:

1] daytime, 9:00 am through 3:00 pm ... up to 300 miles
2] sunrise/sunset up to daytime ... up to 1000 miles
3] night ... up to 3000 miles

4-15. 40 METERS PROPAGATION

40 meters. On the 40 meter band, D-layer absorption is not as severe as on the lower bands, thus daytime skip via the E and F layers is possible covering a radius of approximately 500 miles. At night, reliable worldwide communication via the F2 layer is common.

Atmospheric noise, although still a problem, is not as severe as the lower bands, and often DX signals are sufficient to override existing noise levels. Many consider 40 meters as the lowest-frequency band for year around, reliable DX. During the low points in the solar cycle, Amateurs gravitate toward the lower frequencies and 40 meters is very popular for DXing at night. Phone operators will usually have to operate split as many countries cannot operate above 7.150 MHz.

4-16. 30 METERS PROPAGATION

30 meters. An Amateur band since the 80's, this band has propagation characteristics of both the daytime and nighttime bands. Since D-layer absorption is not significant, communications up to 1900 miles are typical during the daytime, and halfway around the world via the darkness paths. It is usually open via F2 on a 24-hour basis, depending on the MUF. The 30-meter band shows the least propagation variations over the 11-year solar cycle.

Power output on the 30-meter band is limited to a maximum power of 200 Watts PEP, CW, RTTY/Data and Amateurs must avoid interference to the fixed service outside the US. This is a great band to work DX, as many ops do not work CW and the power limits are an equalizer. No big guns or contests allowed here.

4-17. 20 METERS PROPAGATION

20 meters. The 20-meter band is considered as the premier DX Ham band, patrolled with many "Big Gun" stations with huge beams and towers – not for the faint of heart in a pileup. Atmospheric noise is not a serious consideration, even in the summer season. Regardless of the solar cycle, 20 meters can provide at least a few hours of worldwide F2 propagation during the day. During solar maximums, the band will often stay open all night long. Skip distances are great and long path openings are a DXers dream. For example, from San Diego, California to South Africa – the Short Path bearing is 103 degrees (9,908 miles); the long path is 282 degrees (14,972 miles).

20M can have high absorption during summer mid day during high sunspot activity but may stay open all night. During low sunspot activity, 20M tends to become a daytime band. Then there is skewed path where the signal comes from an unexpected heading. Some reports describe a skewed 90 degrees from the true long path heading. During good propagation conditions, three-way communications between widespread continents can often be heard. Example, England to USA to Australia.

4-18. 17 METERS PROPAGATION

17 meters. Propagation on the 17-meter band resembles the 20-meter band conditions, however the effects of changing solar activity are more pronounced. At the peak years of the solar cycle, the band is reliable for daytime and early-evening DXing, and can extend to well after sunset. During intermediate years, the band typically opens during daylight hours and closes shortly after sunset. At solar minimum, 17 meters may be open to the middle and equatorial latitudes, but only for short periods during midday on north to south paths. Power outputs on this band can be in the 100-Watt range with good results. Contests are not allowed on this band.

4-19. 15 METERS PROPAGATION

15 meters. The 15-meter band can be wide open during the peak years of the solar cycle, but somewhat sensitive to changing solar activity. During peak years, this band is excellent for daytime F2-layer DXing, staying open well after sunset. During the low points of the 11-year solar cycle, 15 meters may not open at all except for infrequent north-south trans-equatorial paths. During the intervening years, the band is basically a daytime-only band, closing shortly after sunset. Good propagation makes the 15m band less dependent on transceiver power outputs and many operate here with 100 Watts or less.

4-20. 12 METERS PROPAGATION

12 meters. Combining the best features of the 10- and 15-meter bands it is primarily a daytime band, but may stay open well after sunset during the solar maximum. This band is more conducive to operating at lower power outputs and lower antenna heights, thus making this band very attractive for beginners when it is open. During years of moderate solar activity, 12 meters can open to the low and middle latitudes during the daytime hours, but it seldom remains open after sunset. Periods of low solar activity seldom cause this band to go completely dead, except at higher latitudes. Occasional daytime openings, especially in the lower latitudes, are likely over north-south paths.

The main sporadic-E season on 24 MHz lasts from late spring through summer and short openings may be observed in mid-winter. Contests are not allowed on this band and many old timers use Tri-Band antennas and haven't put up WARC band antennas, so activity here is no way near as heavy as on the 20 meter band. One can hear an African station calling CQ on 12M with no takers. <u>Contests are not allowed on 12M</u>.

4-21. 10 METERS PROPAGATION

10 meters. The 10-meter band is subject to wide swings in propagation characteristics and also has several types of propagation modes.
During solar peaks, long-distance F2 propagation is very exciting where low power stations can produce strong signals halfway around the world and the band is usually open from sunrise to a few hours past sunset.

Propagation tends to follow the sun and for the USA, with Europe coming in the morning, a little later Africa, then the Caribbean, Central and South America with good propagation across the US. Later, the Pacific Islands, then Asia.
There are also seasonal variations, and 10 meters tends to be better near the spring or fall equinox. Summer conditions can be poor even with high solar activity.

Ten meters and 12 meters can be exciting using low power and reasonable antenna heights. Old timers will say 10M can be worked with 5 Watts and a wet noodle, and it is true during the height of the solar cycle. I have worked China by loading my rain gutters. One urban lore is in the old days when a light bulb was used as a dummy load, OT's could work several hundred miles on the light bulb, so they say!

During the solar minimum, there may be no F2 propagation at all. Except for line of sight contacts, the band can be completely dead. In the intermediate years, 10 meters usually opens only to low and trans-equatorial latitudes around noon. During the peak years, signals tend to skip over the close-in states and from California to Wyoming is very difficult. However, Ole Sol makes up for it in a different way – sporadic E propagation when Wyoming can be loud and clear if some of their 1500 hams will come on the air.

Sporadic E propagation is fairly common on 10 m, especially in the months of May through August, although it may appear at any time. Short skip, as it is sometimes called on the HF bands, has little relation to the solar cycle and occurs regardless of F-layer conditions. It provides single hop communication from 190 to 1400 miles and multiple-hop opportunities of 2800 miles and more.

Ten meters is a transitional band sharing some of the propagation characteristics of the VHF bands. Meteor scatter, aurora, auroral-E and trans-equatorial spread-F provide the means of making contacts out to 1400 miles and more.

Beginners may ask why they can only hear one station in a two way QSO, unlike 40M or to some extent 20M. This is explained by skip zones, where you may hear across country but don't have a path to 400 miles away.

4-22. PROPAGATION GRAPHS

Following is a graph that gives an idea of the propagation on the various bands from the USA West Coast to the Cayman Islands in the Caribbean as a function of time. This plot was put together by K9LA for their ZF1A multi-single CQ Worldwide CW contest effort in 1997.
Notice the lower bands 160M, 80, and 40 are night bands with the upper ones occurring primarily during the day. The x-axis of the graph is in 2-hour increments starting with 00 – 02 UTC. Certain assumptions were made to generate the graph such as sunspot numbers, antenna types, etc. Graphs can take this form or in the ARRL format showing MUF, etc. Many of the propagation programs such as IONCAP, WinCap, and Miniprop will plot charts and graphs such as these.

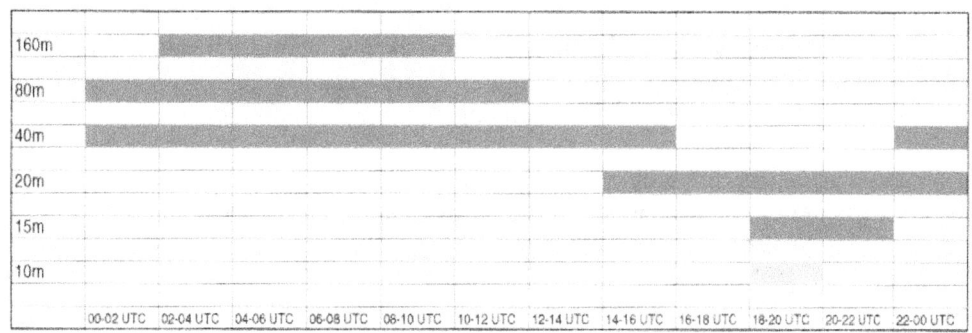

US WEST COAST TO ZF1A

4-23. SIX METERS PROPAGATION

The six-meter band is called the "Magic Band" with good reason, as it can be very unpredictable. This band, like 10M can be very exciting, using low power and reasonable antenna heights and even ground mounted verticals. Excellent results can be obtained using a vertical dipole of wire about 9 feet long, center fed with 50-ohm coax. Of course, beams at 20+ feet and power can give better results. The magic band is subject to just about every propagation mode; tropospheric, meteor scatter, aurora, transequatorial, backscatter, and the biggies - F2 and sporadic E.

Before delving into the details of six meter propagation, be aware that many operators just work what they can hear and are not necessarily aware of the propagation modes in effect. But as inquiring minds want to know, following are basic descriptions of the major propagation modes that the beginner may encounter.

Whole books have been written on the VHF/UHF propagation modes complete with math and charts. The following paragraphs are intended to introduce the reader to the various six meter propagation modes.

Line-of-Sight	Aurora
Sporadic E	Trans-Equatorial Propagation
F2 layer	Tropospheric Bending (Ducting)

Other propagation modes you may hear Hams working are Troposcatter and Ionoscatter. Also there are several more esoteric modes such as meteor scatter, rain scatter, lightning scatter, ice pellet scatter, aircraft scatter, FAI (E-layer field-aligned irregularities), TE (transequatorial field-aligned irregularities), and Moon bounce (EME). For those interested in these modes, some of which are rare or require very elaborate stations, information can be found at URL: http://ac6v.com/propagation.htm#VHF

4-24. 6 METERS LINE OF SIGHT

Line of sight (LOS) distance is dependant on the height of the antennas, antenna gain and directivity, transmitter power, and noise figure of the receivers. Sometimes called the optical distance, LOS is about 30 to 100 miles depending on the fore mentioned items.

Obstructions such as buildings and terrain will affect the reception paths as well. A general guide to transmitter range can be seen at "A simpleton's Guide To Transmitter Range" URL: http://www.artscipub.com/simpleton/simp.range.html

Another general method is the visual distance approximation. Take the square root of the height in feet and that will give the miles from the antenna to the ground. Repeat for the other antenna and add the number of miles. This can be multiplied by about 1.2 to 1.3 for radio waves. For example if the transmitter antenna is 625 feet high and the receiving antenna is 16 feet high, square root of 625 = 25 miles, square root 16 = 4 miles. Add 25+4 = 29 miles for the visual distance. Then multiply this by 1.3 to get 37.7 miles of radio range. This gives the visual distance. And of course, transmitter power and antenna gain, receiver noise figure and antenna gain enter into the calculation.

There are other modes of propagation that will give distances that can be confused with line of sight such as ducting, peak knife edging, bouncing off of objects such as mountains, and tropo propagation.

4-25. 6 METERS SPORADIC-E PROPAGATION

Sporadic E is one of the biggies for six meter operators as it can exhibit very strong long distance communications and it occurs several times a year. Beginners into six meter operation can make contacts across country with just a dipole or vertical and low power rigs (10 to 100 Watts). In the vicinity of the normal E-layer of the ionosphere, patches of abnormally intense ionization can occur. These are capable of reflecting frequencies much higher than those reflected by the regular E or F layers. Sporadic-E was discovered by hams during the 1930s, when the old 5-meter band (56 MHz) produced contacts covering "impossible" distances.

Sporadic E patches usually cover a rather small geographical region, approximately 6 to 100 miles in diameter; they occur at random and are relatively short in duration, typically dissipating within a few hours. Sporadic ionization generally occurs about 60 to 70 miles above the earth, at about the same height as the regular E layer. Hence it is called sporadic-E ionization, or abbreviated Es.

The causes of sporadic-E ionization is not fully known. Since it occurs more often during the hours of daylight, ultra-violet radiation might play some role in its formation.

Occurrences can take place at night, especially during the winter months, thus auroras and meteor trails are often suggested possible sources of ionization. More recent theories suggest that ionization might be caused by wind shear forces associated with rapid wind movements in the ionosphere.

Some researchers connect it with low pressure areas and thunderstorms but there is no evidence of a direct connection with surface weather or regular clouds in the sky. Since the sources of Es ionization are still somewhat a mystery, occurrences cannot be predicted by any reliable means.

The intensely ionized sporadic-E layer makes it possible to communicate over relatively long distances on the 50 MHz amateur radio band, and on some occasions on 144 MHz. The height of the sporadic-E ionization layers limits single hop propagation to an outer distance of approximately 1400 miles. But one hop distances of a few hundred miles are common. During periods of widespread Es ionization, such as several Es patches across country, double hop or multi-hop propagation can extend the range out to distances of approximately 2500 miles. Double hop Sporadic E is not unusual between the east coast of the US and Europe. Multi - hop Sporadic E of three or more hops, is less common but allows for long haul contacts.

When double hop and multihop propagation occurs, the band explodes with signals wall to wall over the spectrum, and stations scramble madly to make as many contacts as they can in the limited time the band may be open.

The six meter aficionados strive for new grid squares and missing states for WAS. If DX out of country appears, it's much like an HF pileup, sometimes with split VFO operation.

Reflections from sporadic-E patches can occur with very low signal loss, so one can experience exceptionally strong signal levels during many openings. However fading and wide signal variations are common. Often it is possible to contact stations well off the great circle path between two stations by means of back and side scatter from a sporadic-E cloud. Es can be very strong with lots of fading and interference. A given event usually affects only small areas of the country at any one time. Beam headings can be just about any direction depending on where the sporadic E patches are located.

Es always affects the lower frequencies first. It builds up from low frequencies to a certain maximum usable frequency (MUF) which may vary widely from minute to minute, and opening to opening.

An Es patch is approximately at the midpoint between the transmitter and receiver, far beyond the visible horizon.

Sometimes the patches remain fairly stationary, but usually they move at speeds up to several hundred miles per hour, more or less in a straight line. This means that the area that can be contacted can vary over a period of time.

In addition there is sporadic E backscatter, turn your beam into the Es patch and listen for weak signals coming from the opposite direction of the strong stations. This usually occurs when the opening is quite strong.

4-26. 6 METERS SPORADIC E SEASONS

Studies over many years for sporadic E propagation have shown peak activity in the summer months with another smaller peak in the winter. Nearly 80% of the yearly totals of Es propagation take place from May through August, with maximums occurring in June or July with June being the more common. Some Es can take place in late April and early September.

A lower but significant occurrence takes place in the month of December. March usually exhibits a definite minimum of Es. However, Es can occur on any day of the year and these are termed off-peak openings.

4-27. 6 METERS HOURLY SPORADIC E OCCURRENCES

Although sporadic E can occur at just about any time, the hour-to-hour variation in Es propagation during the summer months is typically with peaks occurring between 10 A.M. and noon, local time, and again from 6 P.M. to 8 P.M. Es propagation is generally a daytime phenomenon during the summer months, decreasing rapidly after local sundown.

During December, while the peaks occur at about the same local time as they do during the summer months, the P.M. period is well beyond sundown and into the hours of darkness. This December peak may be due, at least in part, to increased meteor activity associated with the Ursids shower which occurs during the middle of this month.

4-28. 6 METERS SPORADIC E GEOGRAPHICAL AREAS

The appearance of sporadic-E is related not only to time of day and to season, but to geographical location. Researchers have identified five distinct geographic zones of sporadic-E occurrence based primarily on seasonal and hourly characteristics. There are also significant variations within the northern temperate zone. Sporadic-E ionization occurs most often in the western Pacific, China and South-east Asia, and least often over the north Atlantic and adjacent portions of the north-eastern North America. In the US, E-skip is nearly twice as common over the South-west as over the North-east. Peak times for sporadic-E in the rest of the world vary considerably.

Like that of the northern temperate zone, the major sporadic-E season in the southern temperate zone occurs from late spring to early summer (mid-November to mid-February in the southern hemisphere). In the equatorial zone sporadic-E is nearly a constant phenomenon of the 8-hour period centered at noon regardless of season, but it is rare any other times. In the equatorial zones, sporadic-E is least likely to appear at noon, but it appears more than half the time in the 1800 to 2400 period with little variation throughout the year.

4-29. 6 METERS SPORADIC E SOLAR CYCLE VARIATION

The data for sporadic E over the years does not exhibit a clear cut correlation with the solar cycle. In some peaks of the 11 year solar cycle – the sporadic E occurrences were low, but in other peak years the Es was high.

4-30. 6 METERS SPORADIC E INDICATORS

There are several ways to detect sporadic E openings:

1) Monitor the lower TV channels (not cable). Indications are bars and interference patterns and even a complete take over of the channel by a station hundreds of miles away.

2) Check the beacon range 50.0 to 50.1 MHz

3) Check the DX packet clusters – command is sh/dx/21 50   This will show the last 21 (page full) of the 6M DX spots.

4) Monitor 50.125 MHz

5) Check the Propagation and Auroa reports. High sun activity may signal openings.

4-31. 6 METERS SPORADIC E MAPPING

After a while when working sporadic E skip, a mental picture forms as to where the ES patch is located. For example when working the northwest states from San Diego, particular grid squares of Washington and Oregon are predominant indicating that the patch is between the two areas. If stations in Idaho are coming in as well as grid squares across Washington, it would indicate a wider patch to the northwest. Indeed some dedicated six meter aficionados plot their contacts on a map in an effort to locate the Es patch. As patches can move rapidly, one can get a sense of the patch speed.

4-32. 6 METERS F2 SKIP

This is the other biggie for six meter operators but is only experienced at the solar cycle maximums (11 year cycles). It is the most common long distance propagation mode at HF and can also result in some tremendous DX openings on six meters. Hops are typically over 2000 miles in range. Over the years, observations have shown that for the higher latitudes, with conditions of very high MUF, the best months for 6M F2 skip centers around December.

In the last part of 2001 and early 2002, the east coast of the USA was routinely working into Europe, while west coast USA stations were working into the Caribbean and sometimes into the Pacific and Japan.

Peak times tend to be around each equinox (Spring and then Autumn).In addition to the favorable months, check for high solar activity which can result in propagation above what might be expected from the MUF. A good source for checking openings is the DX Packet Clusters or just set your receiver for scan mode around 50.110 MHz. Also monitoring 28.885 MHz will find stations with six meter reports and requests for possible contacts.

During periods of high solar activity, the Solar Flux will typically be in the two to three hundred region, sometimes greater. Good 6M conditions are generally, but not always, associated with a high solar flux and a low A index. That is a flux above about 200 and an A index below 8 units. A high A index, say 30 upwards, would indicate the possibility of an aurora. The higher the figure, the more likely an aurora will take place. F2 activity is greatest in years of peak sunspot activity.

As of the date of this publication, we are well beyond the 11 year solar cycle maximum for cycle 23 with the minimum likely to occur sometime in 2006. The next maximum is predicted to be about the year 2010 for solar cycle 24. So one can forget about F layer skip for several years.

But sporadic E offers several opportunities to work six meter DX, but the openings can be days, weeks, or even months with little or no activity. Checking the DX packet clusters will be a method to determine when the band openings occur. During seasons of expected activity, scan the band around 50.110 MHz and 50.125 MHz.

4-33. 6 METERS BEACONS

Six-meter operators do a lot of waiting, because of the unpredictable nature of the band. To help determine when the band is open, six-meter gurus around the world have installed a fairly extensive suite of beacons on the air. In the U.S., beacons occupy the region between 50.060 and 50.080 MHz. Radio Beacons are the lighthouses of the airways. Usually unattended, they send out a radio signal periodically identifying their callsign. From these one can determine propagation conditions around the world. A list of radio beacons is available on the Internet. See URL: http://ac6v.com/beacons.htm

4-34. 6 METERS F2 BACKSCATTER

Backscatter propagation generally occurs when the maximum usable frequency (MUF) rises above 30MHz. When radio waves reach the ionosphere, they are reflected back towards the earth's surface at a larger detectable angle than usual. Backscatter signals reflect at a very sharp angle back toward the transmitting station and can cover the range of the usual ground wave and may be heard out to 1500 miles from the transmitting station. These are characterized with a distinctive hollow sound and exhibit very little fading, although they are usually much weaker than normal reflected waves. During periods of low solar flux, it requires a directional antenna to produce readable signals. During periods of very high solar flux, stations using low power and a vertical antenna may be successful using backscatter. In some instances, pointing the beam 180 degrees away from the desired station can achieve the desired contacts.

4-35. 6 METERS AURORA

Of all the ionospheric propagation modes, aurora is probably the one most observed on the high VHF bands but occasionally on 50 MHz. Aurora occurs at the mid latitudes and is caused by intense geomagnetic activity such as solar flares. TV is often heard via aurora, with heavy distortion and interference from other co-channel services.

Aurora propagation occurs when high levels of charged particles arrive at earth forming an auroral E layer at about 70 miles high. Radio waves from the HF band all the way up to the UHF band can be propagated by aurora propagation. An auroral signal is easily recognized as a burbling sound or under-water-sounding modulation and heavy fading is typical. Because of sudden phase shifts, narrow band modes such as CW and digital are the most reliable modes for DX contacts.

Auroral activity is most likely around the equinoxes, but may happen at any time when the proper solar disturbances occur. Heavy auroral activity can also induce regular Es during the following day or two and at 27-day intervals coinciding with the rotation of the sun.

Southern stations often experience auroral signals, but it becomes less and less frequent at more northerly latitudes. WWV advises of magnetic disturbances and auroral activity periodically; you can also spot auroral conditions from short-wave blackouts and heavily distorted signals on the 20 meter band.

Aurora has distances typically in the 350 to 650 mile range, but with no specific lower or upper limit. Typically stations in the northern latitudes rotate their beams to the north for signals to bounce of the aurora and return to the south, but the aurora signal may not be coming from the direction you would expect as it is bouncing off the auroral curtain, which may be as much as 45 to 90 degrees away from the direction of the station.

4-36.   6 METERS TRANS - EQUATORIAL PROPAGATION.(TEP)

TEP Trans-equatorial Propagation is the result of F layer ionization over the geomagnetic equator. Note that the geomagnetic equator is not the same as the geographical equator; it is defined as the imaginary great circle on the earth's surface formed by the intersection of a plane passing through the earth's center perpendicular to the axis connecting the north and south magnetic poles.

Contacts are made from opposite sides of the equator and good contacts have been possible even through the minimum of the solar cycle. During the solar maximum, both the north and south paths are extended well past the normal temperate zones. TEP has a distinctive flutter since the MUF changes very rapidly. SSB may be almost unreadable, so CW is used frequently. Distances of between 3,000 and 5,000 miles are not unusual, and even stations not strictly on the other side of the equator may come in by this method, if they're beyond 1,800 miles. TEP occurs during the spring and fall months, being best at higher sunspot numbers.

The southern part of the USA experiences a number of TEP openings to South America. The stations are generally about equidistant either side of the magnetic equator. Given exceptional luck, an Es opening linked into this mode can make it available to more northern stations. This mode has bad flutter fading and a touch of the auroral spectrum spreading. One of the first signs of improving sunspot conditions is an increase in this type of propagation.

4-37.   6 METERS TROPOSCATTER

Troposcatter has been used commercially and by some military sites for many years. The optical line of sight communication distance is about 30 miles to 100 miles depending on altitude, power, antenna gain and other factors. But you will hear stations working troposcatter out to distances of 500 miles or so. Usually these are stations with high beams and running power and perhaps a good hilltop location.

Troposcatter is scattering of radio waves in the troposphere caused by irregularities in the atmosphere. These irregularities are believed to be small changes in humidity, temperature and pressure.

The troposphere is the atmosphere below an upper limit called the tropopause, which is at a height of about 6 miles. Above the tropopause, the temperature is constant, so there is no humidity and few irregularities to scatter radio signals.

4-38.   6 METERS IONOSCATTER

Ionoscatter (IFS) is similar to troposcatter but is caused by scattering from irregularities in the ionosphere at about 4 to 5 miles of height. This may give a range of up to 1,200 miles, if you have 500 Watts into a 12 dB antenna. Ionoscatter has a daily variation of signal strength, usually noon is the best time at the path midpoint. During the daytime, the height of the scattering media is somewhat lower, so the range may be lower but with better signal strength.

Path loss increases below some 750 miles due to increasing scatter angle, so a path from 750 to 900 miles is considered favorable for ionoscatter. Signals are weak, and it takes good beams, height, and power, but it is very reliable.

4-39. 6 METERS TROPO (TROPOSPHERICBENDING)

Tropo is another major propagation mode and is dependent on weather conditions in the troposphere. There is no minimum distance for tropo. Depending on your equipment, you may notice tropo improvement on stations as close as 50 miles but distances in the range up to 600 miles are not uncommon. A duct may appear and vanish in little over an hour, or may last for days. Tropo is the steadiest of any propagation; it seldom has rapid fading, but may fade slowly in and out. Weak tropo in the range slightly beyond that normally received is often called extended ground wave.

Temperature inversions can result in propagation over extremely long distances, for example from California to Hawaii on rare occasion. These inversions allow ducting or channeling of radio signals to occur between two air masses of different temperatures.

The coastal areas are prime candidates for ducting as large temperature variations typically occur in these areas. Spring and fall months seem to be the best, when there is a fairly wide temperature variation between day and night. Tropo builds up quickly after sunrise but tends to "burn off" during the hot afternoon hours; it may fade back in after sunset from the same area seen in the morning.

4-40. 6 METERS METEOR SCATTER (MS)

There are several ways to work meteor scatter. The long time favorite is high speed CW (HSCW). This is at speeds of up to 6000 letters per minute and even faster. This translates to 1200 words per minute! At this speed, a 400 millisecond meteor burst will carry about 40 letters which is sufficient for communications. Distance ranges between 500 and 1300 miles. More information on HSCW can be found at URL: http://www.nitehawk.com/rasmit/hsms-intro1.html

The latest rage in meteor scatter is an application called WSJT. There are two operating modes FSK441 and JT43. The FSK441 mode is for operating meteor scatter but unlike the older High Speed CW, the computer now decodes the pings and displays them in print in the receive window. The program provides a method of saving wave files and printing the decoded text to a text file if one wishes to use these features. Distances between station and direction pointing are provided on screen.

Be aware that working meteor scatter has a format, signal reports, and jargon all its own. These can be found at the following URL's:

http://ac6v.com/opmodes.htm#MS

Another excellent article on meteor scatter can be found at URL:
http://www.uksmg.org/deadband.htm

4-41. 6 METERS PROPAGATION MODES COMPARISON CHART

Mode	Times/Seasons	Typical Equipment	Typical Range
Line of sight	Any time	10W and small antenna	0-30 miles
Sporadic E	Sporadic but has typical seasonal times	10W and small antenna	600 to -1,500 miles single hop; 2500+ miles double hop or triple hop
F2	Depends on SFI, A and K solar indexes	10W and small antenna	at max sunspots up to 10,000 miles
Aurora	Sporadic	10-100 W and a beam	250-1200 miles
Meteor scatter	Any time, but especially during the seasonal meteor showers	10-100 W and a beam	500- 1500 miles, longer during meteor showers
FAI*	Sporadic	100 W and a beam	No Data
TEP*	Sporadic	100 W and a beam	Up to 3,000 – 5,000 miles
Troposcatter	Any time	100+ W and high beam	30 to 500 miles
Ionoscatter	Any time	500+ W and 12dB+ beam	600 to 1200 miles
EME	Dependent on moon transit times	1+ kW 18 dB beam	Up to 10,000 miles

NOTES

CHAPTER V	WORKING HF DX

5-1. CHAPTER CONTENTS

NOTE: Throughout this section, reference is made to the Callor – that is the one running the pileup. The callee is the poor guy trying to get a contact.

5-2. EXPLODING THE BIG GUN MYTH

Over and over again the budding DXer will exclaim, "I'd like to work DX, but there is no way I can compete with the Kilowatts, beams and the amps". Certainly the big guns have an advantage but be assured you can work DX with a modest station. Hopefully this book will give you a head start in that direction. Here are some examples. All are DXers that I know:

George has a DX-77 Vertical, runs 100 Watts and has worked over 280 countries in this sunspot cycle, all CW.

Paul has an FT-100 and ATS Whip Mobile -- worked 50 countries in just 3 months. CW & SSB

Rod has a Cushcraft R-7000 Vertical and 600 Watt linear and just made Honor Roll, 2/3 phone, 1/3 CW.

Bob has an AEA Isoloop in his attic and has worked DXCC 100+.
Ron operates QRP with a Butternut Vertical, has 120 countries confirmed. All phone!

In case you have been incorrectly informed that this current sunspot cycle is a loser. Joe, W1JR, states, "Just a note to tell you DX this year is doing well and possibly ahead of all years. Each year I try to see how many DXCC entities I can work. In a good year I can make about 280-290. Hence, on January 1, 2001 I started anew AGAIN. I guess I never learn! Yesterday (May 3rd) I worked XU7 for DXCC #270, all in just over 4 months. This is the earliest in the year that I have ever worked 270 entities. I figure I have missed about 10 others. I do keep a normal sleep pattern and don't catch them all! The DX is there. You just have to listen and work them!"

With knowledge of propagation, operating techniques, and some patience, anyone can work DXCC quite easily. Some have done it in a contest weekend. Also it is not true that it is all "QSL ur 59(9) 73". Long conversations can be had with many DX stations in Europe, South America, Asia, Caribbean, Australia, New Zealand, and many other parts of the world. Many DX stations don't like the "send me a QSL card" QSO, and they can be quite chatty outside of a pileup.

5-3. FINDING DX

If you are new to HF DXing, the *USA Band Plans, Modes Of Operation Allowed, and Power Limitations* are discussed at the end of this chapter. Also listed are the *Considerate Operator's Guide* and *Calling Frequencies* including DX Windows and common DX Frequencies.

In the old days of DX, the old adage of listen, listen, listen was the way to find new DX countries and perhaps this is still the best advice, but technology is with us and today there are other major sources of finding DX. Old timers will lament "It ain't like it used to be, we had to find our own DX, not some new fangled, Spoon-Fed, Packet Cluster, or getting DX info off the Internet". But that was then, not today, so let's look at the old adages and the new techniques and tools available to today's DXer.

- Listen, Listen, Listen
- Propagation Reports and Beacons
- VHF DX Packet Clusters
- Telnet DX Packet Clusters
- Internet spotting networks such as OH2AQ and DX Central
- DX News Letters (on-line and printed)
- Contests, both CW and SSB
- Islands On The Air (IOTA)
- DX Clubs
- Working DX Nets

5-4. LISTEN, LISTEN, LISTEN

One can spend hours and hours tuning the bands for DX and for the beginner with a low country count this can net a bushel of DX stations per day. In addition your ears will become accustomed to the difference between domestic and DX signals, not necessarily signal strength, but a signal from far off can take on a different characteristic, fading, watery, polar flutter, and other phenomenon. For phone, you will start to pick up on accents and can often tell if it is a DX station just from their manner of speech. Australians are a good example.

Many beginners will start by tuning the whole band, but not finding much or any DX in certain edges, make the mistake of just tuning the "DX windows". Tune the whole band. Many a time DX is heard at some "non DX" frequency such as 14.075, 14.348, 21.445, 28.700 MHz. And since many of the new rigs have 6 Meters and DXpeditions are using this band – a six-meter rig will also net new DX stations. RTTY is becoming popular with DXpeditions so best to have this mode setup and ready. Some DXpeditions have even worked 10 Meter FM.

By listening and tuning, you can work many of the more common countries without the aid of the DX Packet Clusters and may get lucky and catch a rare one. Remember, someone on the packet cluster had to listen and tune to find them – might as well be you. As you tune, the signal strength, fading, accents, flutter and other characteristics of the signal will soon allow you to differentiate between DX and domestic long before the station identifies. After awhile you will begin to recognize individual DX and domestic operators just from their voice.

5-5. THE PILEUP EXCHANGE

A typical pileup exchange between the DX station and the callers: "This is DX11DX from Charcoal Island QRZ".

The caller simply responds with his/her callsign phonetically "Alpha Charlie Six Victor". No need to give the DX callsign. Use standard NATO phonetics or the DXer phonetics in Appendix A4, e.g., "America Canada Six Victoria". DX11DX responds with "AC6V 59 (QSL)" or "AC6V 599" on CW.

AC6V acknowledges with "Thanks 59 ur 59 AC6V" or "QSL 59 – ur 59 from Alpha Charlie Six Victor", or on CW "QSL 599, ur 599 de AC6V". Some just give their CW report "599 de AC6V". On CW the report numbers are frequently abbreviated, with the letter N used for nine thus 5NN is sent. Short and sweet as the DX station wants to move on and achieve as many contacts as possible. Some callers insist on giving their name and QTH as part of the report, this is usually wasting time; only give this info if the DX asks for it.

There is a rhythm to a pileup or should be. How many times should you give your callsign? Listen to how the DX station is picking out the callers, many DX stations can respond rapidly to calls in a pile-up, picking up one on the first round of call-ins. Others may not respond until 2 or 3 calls. This will determine whether you give your call once or more times. Inexperienced pileup management results in a free-for-all and you will have to listen to see how the DX station is handling things, if at all.

One might ask about 59(9) reports when in fact the DX is weak, maybe a 3 by 4. Well many DX stations and particularly DXpeditions and contest stations don't want to waste time writing or typing a bunch of reports, so they draw a line through the report column of their log, or program the computer for all 59(9) reports.

Lass Too – or calling with only the last two letters of your call may be against the FCC rules, never the less some operations take on this characteristic. But it has been suggested that as long as you ID within the rule time limits, it doesn't matter how you call, even with two letters, so long as it is not deceptive, of course. The problem is where a station only ID's with their suffix, never getting the opportunity to give the full call within the time limit. But why not give your full call when first entering the pileup, then followed by calls of "lass two" until the 10 minute rule applies??? Also imagine 3000 callees giving their full call sign, now picture those same 3000 calling with just the last two letters.

Listening to pileups, one comes across the DX operator who can not seem to get an entire call like Double Ewe Ahh Six Double Ewe Tee Oh and consistently picks out only a letter or two of the call. So folks start sending the last two and the DX op favors these. Other DX stations will state last two only. Many of the DX Nets -- get a list by calling for last two -- and the netters get into the habit of this goofy procedure.

I have heard numerous tail-enders and quick callers come in with the last two and son-of-a-gun if the DX doesn't pick up on it. The good operators ignore all this and insist on full calls and do not take tail-enders before he/she QRZ's again. A sharp callee will pick up on what is the best way to get thru --- and if the DX is favoring last two and tail-enders with last two – then so be it.

Some DX stations will ignore callers using the "last two", so listen for a bit and you can determine how the callor is picking out stations. When the DX contacts you with "last two", they will need your full callsign and this somewhat makes things legal. After all, in a pileup, who knows what you said before you announced your "last two". This works for DX operators who can't seem to get a complete call. You'll hear "the WD6 – go ahead", this is a tipoff how the DX op is perceiving the pileup. Most good DX operators discourage the practice as it slows down the Q-rate.

In a contest, the report might include a serial number 59(9) 307, or a zone report 59(9)06. CW contesters almost always use 5NN as an abbreviated form of 599.

After working several stations, the DX station usually announces QSL via (CBA – call book address), (Burro - bureau), (my manager W10XYX). They also will announce their location if it is not obvious from their callsign prefix. In severe pileups, the DX station may opt to work split (up 5) or by selective calling (W6's Only – California), EU only – Europe).

5-6. THE ANATOMY OF A PILEUP

If you encounter a pileup it is sometimes puzzling as to what's happening. As an example let's say you encounter a pileup and after listening for a bit you can't hear a DX station. There are four possibilities:

1. The DX station is simplex - listening and transmitting on the same frequency. e.g., 14.195 MHz.

2. The DX station is operating split but only listening on one frequency e.g., transmitting 14.195 listening 14.205 MHz.

3. The DX station is operating split but listening on a range of frequencies e.g., transmitting 14.195 listening 14.200 to 14.210.

4. The DX station is operating split but listening on a discrete multiples of frequencies e.g., transmitting 14.195 listening 14.200, 14.205, 14.210, 14.215. This is rare and peculiar to certain operators.

5. In 1 above - the DX is operating simplex, listen for a while and determine if you can hear the DX, you may not have propagation to that point in the world but the other callers do. Check the packet clusters, chances are you can determine who is on from the spots. Don't ask on frequency, this just adds to the confusion. Another tip off is there are no cops saying he's working split.

In 2 above - the DX is operating split and listening for callers on a single frequency. Example the DX is on 14.195 and he/she is listening 14.200 MHz. Since split operations are usually listening up, try tuning downwards to find the DX station. Usually (unfortunately) a cop or two is guarding the frequency telling others "he is listening up" "he is split", "wrong VFO", and insults as well. A check of the DX Packet Cluster usually has the DX call and whether they are split or simplex.

In 3 above – this is the usual split operation and affords quite a few frequencies for folks to call on. If there are cops on the DX frequency, it is a sure bet the operation is split, however sometimes a cop comes on a simplex operation, warning that the DX is working by district or country now.

In 4 above – this can be confusing, but use your TF Set control (tune the split range) and determine if there is a pattern to the DX station listening habits. Again tune down (usually) to find the DX station.

There are three major categories of pileups, DX Station, DX Visitor, and the DXpedition. The DX Station is resident in the country and operates at his/her leisure, either retired or working with the attendant operating hours.

The visitor may be on holiday or temporarily working in the country, and of course the DXpedition is there for full time DXing looking for as many contacts (Q's) as possible.

As the packet spots increase, so do the number of stations who call. The goal of the DX operator working the pileup is to efficiently work as many stations as possible. But the DX station needs to control the pileup, and an out of control pileup is one of the biggest causes of frustration on the HF bands.

One of the big problems with pileups is that many of the stations calling can't hear the DX because of the QRM caused by other stations calling also. The simplest and most efficient system for solving this problem is the split operation - where the DX station transmits on one frequency and listens on a different frequency, or over a range of different frequencies. More on this later.

The indigenous DX Station may be a rag chewer or award hunter (county, state, 10/10) and many times presents an opportunity to carry on a conversation, where the visitor usually likes to run quick ur 59(9) type of contacts. The DXpedition is almost strictly a high Q-rate operation. When any of these generate a pileup, there are several methods of managing the howling mob.

Sharp operators with strong signals may opt to operate simplex; that is, they call on a frequency and listen on the same frequency. These folks are quite adept at picking out call signs and can achieve upwards to 300 contacts per hour (5 a minute) on SSB and up to 200 contacts per hour (3 a minute) on CW.
Calling technique can be:
Calling anyone, anywhere
Calling by continent
Calling by country
Calling by propagation patterns
Calling by district, W1's, W2's, etc.
Non-sequential method
Taking their own list
Specialized – YL's, QRP, mobile, age, Ham clubs, etc.

The problem with calling anyone – anywhere is that the pileup can get huge and ugly very quickly and propagation may not favor you and it can be tough to get through.

Calling by continent or country thins out the pileup but if carried on too long results in loss of propagation and folks get irritated.

In the propagation method, the DX works the areas where propagation is likely to fade before the other areas. For example, the DX may work W1's, 2's, and 3's first, then moving west as the propagation changes. Indeed some savvy DX stations call for each district to ascertain the propagation. Even so, the whiners interrupt with 'When are you gonna work the sixes ?'

Calling by district can be unfair but tis a fact of life, for example, the W6 population is around 102,900, but W3's number 37,200. JA1's are perhaps six times the number of the 6th district. One noted DXer suggests a better way might be for working the USA, is to combine districts 1,2 and 3, 8 and 9, and sometimes 5 and 0, or 6 and 7, and of course 4's all alone.

The non-sequential method calls districts out of order – for example the DX calls W8's and W9's before they work W4's because the DX operator is knowledgeable about the propagation patterns and they want to work as many from each propagation area as possible.

Some DX stations like to operate by taking a list, then working only one caller at a time and refusing all others until the next list. This is commonly used by DX stations that want to chat a little such as name, location, etc.

Calling specialized (YL's, QRP, etc.) gives some advantage if you fall into this category, maybe unfair but it is the DX who is running the show.

Then there is the DX station that is working W4's only – but answers a W2. Go figure. Results in total chaos. And the one who doesn't announce when he/she is going QRT – they get to your district and QRT, some fun.

As for the mob calling in, there are several breeds. The ones who follow the DX station's instructions, then the ones who don't know what is going on, and the ones who figure it is all fair in love, war, and DXing.

The ones who don't get it can be heard on the DX frequency asking, "Who's the DX, What's his call, where is he listening, etc." Cops (guys who already worked em) will offer everything from advice to insults and do little but slow down the Q rate. There is really no excuse for anyone to ask on frequency what is going on, checking the packet clusters or telnet will usually reveal all, including QSL information. For those with no packet or Internet, they can always go below the DX calling frequency and ask, "what's going on up band ?"

For the pileup bullies, they call out of turn, tail end, and this breed proclaims "Forget about good manners, DXCC isn't about good manners, it's about who gets to the top of the mountain first". Some DX stations will warn the bullies and dummies that are not following the protocol and start a blacklist so will not work the offenders.

Then there are the whiners covering up the DX frequency with "When ya gonna work sixes", "QRP QRP" "We are losing propagation" "Tell him to standby for mobiles" and other begging techniques. All this causes a slowdown of the pileup and is counter productive.

Frequency cops can be a pain; they often create more interference than the occasional on-frequency caller. Usually the cop(s) come on right after the errant one - usually right on top of the DX station. And cops love to beat up on the errant caller. All this slows things down, and helps very little. It is better to just let it go and see if the errant one gets the picture, if not, a simple "Listening Up" should suffice.

Another no no is tuning up on the DX frequency or even the split range. Why is anyone's guess? All in the game.

There is a rhythm to a pileup (or should be), a good DX operator will foster this by giving his call frequently, stating listening this frequency, up (200 to 210) and is consistent with the turnover - either QRZ, his call, your call QSL, another, etc., and use this all the time as it maximizes the Q-rate and the callers soon get into the swing of it.

Some pileups go as smooth as silk because the DX station establishes a consistent pattern. Some DX stations acknowledge the caller by callsign to be sure they got all in the log OK, thereby eliminating dupes and insurance contacts. My favorite is the DX station that signals the turn over with "DX11DX, QRZ -- or DX11DX 200 to 210". At the very least they should identify once every minute or so. QSL info is usually given at longer intervals or not at all if it is a well-published operation – posted in all the DX bulletins and the web.

Operating split. When the DX can't handle a pileup or it gets too hairy, they will operate split. See paragraph 5-7.

Insurance contacts are definitely in order. A DX operator can call you correctly, but write something different in the log. Maybe on CW, the DX worked AC6ST, not AC6V, and numerous other anomalies can occur. Ideally, work em on a different band, but on the same band if you have to, just be sure you are in the log. A DXpedition may not occur for many years to that DXCC entity. DX ops hate dupes, but if there is any doubt, call again if it is rare rare DX.

Nothing is as shattering as to spend hours calling the rare DX and months later getting the "Not in the log" report – worse than a Dear John letter to the true blue DXer!! But be aware that some DX operators actually publish rules in the DX newsletters or a web page – NO DUPES. Be certain that your first contact is in the log, even if you have to ask for clarification on the initial contact.

DXpeditions afford a great opportunity to get that new country. In the recent D68C operation, they worked over 168,000 contacts with over 45,000 DXers.

5-7.WORKING PILEUPS

Are pileups just madness and insanity? Well, no more than football fans at the big game, or teenie boppers over the latest rage singer, or perhaps YOU at the bowling league playoffs. But there are ways to deal with the chaos. Lets look at some "Little Gun" techniques.

For the Little Gun stations running 100 Watts and a minimum antenna (vertical, low dipole), CW DXing will probably yield better results. First with the advent of the 5 WPM Extras and Generals not very many of these will work CW pileups. Also some percentage of the 13 wpm and 20 wpm licensees do not work CW either, so overall the competition is usually much less than on phone. RTTY offers even less competition. Toward the end of a DXpedition, it is much easier to work them on CW than on phone as many of the Big Guns have made their contacts.

A well-run pileup will have the DX QRZ and he/she immediately picks out a caller, thus you should transmit immediately. But if the DX operator hesitates waiting for another round of calling, unless you are one of the biggest guns around, don't bother transmitting your callsign as soon as the DX announces QRZ. Everyone instantly keys up and the strong survive. Wait a few seconds for the pileup to subside and then transmit your call. I say to myself "Whiskey Six American Dog" then scream away. But this doesn't work and should not be used if the DX can pick out a call immediately after the initial onrush. Get in the swing of the pileup cadence and it will help facilitate a faster Q-rate.

Be sure the DX station has your correct call, which can be difficult in a big pileup. If in doubt, log the calls of the stations worked before and after yours so that you have some proof you made the contact and the call in the log was recorded incorrectly.

Another trick for working a non-split pileup during a lull is to say your call quickly without phonetics. AH CEE SIX VEE has pulled thru on many occasions. Also stretching out your call under weak signal conditions may help, e.g., AHHH CEEE SIXX VEEEE.

On CW, two strategies can be used, the first for fairly strong non-polar DX stations and the second when signals are weak, particularly with polar flutter. When signals are strong and clear, send at the same speed or slightly faster than the DX station.

If the DX station is sending manually (no memory keyer), listen for a bit and you will be able to tell, as it is very difficult to send perfect code by hand. In this case sending faster than the DX station is a good bet as many ops can copy faster than they can send manually.

A typical "calling window" might be 2 or 3 seconds, so adjust your sending speed for fast responses. CW DX stations and DXpeditions might send any where from 25 to 40 wpm, so get into the rhythm of the sending receiving pattern and this will help you get through the pileups.

In the case of weak signals with polar flutter, the hollow, fluttery sound can make it difficult to copy (on both ends), so contrary to many callers, slow down your sending speed considerably, perhaps to 12 to 15 wpm. A lot depends on the DX operator, but slowing down on weak fluttery signals seems to work more often than not.

Know how to "zero beat" a CW signal. Many stations have very narrow filters and you want to be in their bandpass. Refer to the operating manual for your radio. However one CW pro advises, "The assumption that replying at zero beat is highly desirable is incorrect. Good and competitive CW operators know that calling as much as 200 Hz off frequency is usually much more productive in pileup situations. The XIT control is useful for this. A disadvantage of razor-thin filter selectivity is that this very useful strategy does not work well. A good operator uses his "ear" as an adaptive filter and lets the receiver cover a few hundred HZ."

If after transmitting, you hear a quiet pause on the frequency - key up - that is your green light to jump in. Many times the DX station can't pull anyone out of the pileup. A quiet period just after the initial roar is your queue. The quiet pause is not as common these days – but does occur.

Tail ending is another controversial technique. For tail ending in a simplex pileup, nothing is more disconcerting than to behave yourself in a pileup (waiting for the QRZ), and then have some clown throw in his call on top of, or immediately after the responding station. If the DX station works the interloper – then it is fair game.

I don't tail end unless it is evident the DX will accept tail-enders and after all the DX station runs the show. Tail ending of a different type is also used in a split operation, to be discussed under split operations.

Don't give up. Amazing how many beginning DXers call a half a dozen times and give up. This is serious business folks, if you want that S79 – hang in thar. My first Seychelles contact took me an hour and a half. When he closed down, a W9 asked me what possessed me to call for an hour plus. I coolly replied "because it's there" ala Sir Edmund Hilliary. Went over his head I'm sure (maybe mine too huh?).

If you encounter a DX station in a semi–pileup doing some rag chewing, don't play little gun and give up, especially if the DX is chatty, they may ask for "West Coast", your state (helps if you are in Wyoming), or technical information.
I recall an A61 station who wanted to know how to increase the power out of his TS-850 in order to drive his linear to full power – only one station came back with the info.

If you are in a rare state, add your state name at the end such as KL7DX in Alaska. Amazing how those weak Wyoming stations get thru. Numerous times in a pileup you can hear the DX say "The station in Rhode Island - go ahead". Also useful is to holler "west coast" or "California" if the pileup is mainly elsewhere.

If the DX calls for QRP stations, turn off the linear, turn your RF power to minimum and jump in. Many a DX operation will call for YL's only. I have heard a DX station call by age, for under 16 only. If you have an oddball prefix, the prefix hunters will call you! Oddball prefixes will also have the DX call you by the bushel during the CQ WPX contest.

For USA callers, calling CQ DX can be disappointing, but on occasion will work. Many have fallen off the operating chair when a JY or 5N replied, but this is the exception for sure. Calling CQ DX works especially well for Big Gun stations "when the band goes out" or is just opening. Get that radiation angle down low! Try calling CQ DX, No QSLs, Only QSOs!!

When a DX station announces they will QRT or QSY, throw in your call, especially if you have met or know the operator. P40V (AI6V) always comes back to me!

Strange how many DX stations announce QRT then come back in a few minutes (potty break?) Hang in there.

The DX says "Back in 10 minutes", maybe -- but better tune up and down the band, they may not want to come back to this zoo of a pileup. Happens often. Put this frequency in memory and tune around or check the other bands.

Some DX Stations go silent and let the pileup die down and then listen for the little guns and make a list, coming back on and working their little gun list.

Tune the bands after they "go out" because they sometimes come back up. One night a fairly rare African station called CQ on 17M late at night - no one answered - except one W6.

Listen for tuning carriers. Hmm - but many times it means there's something interesting to work especially if the carrier has that DX sound and is on one of the common DX calling frequencies. As soon as the tuner-upper is through, announce your callsign, you might be surprised who is on frequency.

ESP contacts. Maybe Expected SP. The faint of heart will exclaim, "I can barely hear the DX station". Well that is good enough. Example, it is late late on 20M, the east coast is asleep, and the west coast is hearing watery weak polar flutter signals at best. R1MVI is CQing – maybe every 3rd syllable is understandable – no one responds. So what if it takes six overs to get the reports across – jump in. On several memorable occasions, the Little Gun with good ears has been the last, or one of the last, to work the DX before they shut down or went to another band.

The human brain seems to be capable of a certain amount of weak signal processing. When listening to a very weak signal for a long period of time, the brain seems to be able to filter out the background noise, and the weak signal becomes discernable.

Through this book, I've avoided the brag tape, as I'm sure you would be bored with the author's exploits. However, let me tell the story of my Mapelo DXing, only as a matter of how to snag a rare weak DX station. I had worked Mapelo before, but the ARRL didn't grant credit. Pedro HK3JJH was on Mapelo for several weeks, but in Southern California his signal was extremely weak on any band at any time. Using the DX packet cluster, I listened on all bands and kept a log of results. The only band I could even hear a whisper of a signal was on 20M around 0200 to 0330Z. I tuned in every night, never called because I couldn't have heard my call sign even if he came back to me, but I could hear the whole world working him.

After two weeks, the SFI, A and K were getting favorable and I began to catch syllables, still I didn't call. By this time the pileups were diminished, he had worked so many and many couldn't hear Mapelo on the west coast, and by now he was working simplex.

Then the opportunity came, I could hear whole words. Using RF Gain down, AF Gain up, IF Bandwidth at 3 kHz, I had a copy using Sony Hi Fi headphones. But a strong station up band on 20 M was clobbering Pedro's signal. After an hour and a half – everything fell into place, his signal was as good as it was going to get, the east coast was asleep, the up-band station went QRT and no one answered his QRZed. I called for the first time ever, and through the noise, the mists of ESP, luck, and dogged determination, I heard Alpha Charlie Six Victor, Five and Nine!!! Good Golly, Ms Molly, DX IS!!! More DX Tips in Chapter 7, DX Secrets.

5-8. OPERATING SPLIT

With the advent of the DX Packet Clusters, pileups can develop in a matter of a few minutes (seconds). Many DX operators and certainly DXpeditions desire to make as many contacts as possible and frequently "operate split". For beginners, it's somewhat of a puzzle as to what's going on and every big split pileup has dozens of stations not aware of what's happening.

When a massive pileup occurs, the DX station may have trouble picking out the callers, but worse the callers transmit indiscriminately over the top of the DX station and each other, covering up the DX station who may be weak anyway. Bedlam ensues; shouts of lid, cursing, and the radio police make it worse, etc.

The most efficient system is for the DX to listen at a different frequency, that is the DX station transmits on one frequency (e.g., 14.195) and listens on a different frequency, or over a range of different frequencies, for example, "Listening 14.200 to 14.210". The object of all this is to move the callers (the pileup) far enough away from the DX station so that the pileup doesn't clobber the DX transmissions. Thus in our example, your receiver is set to 14.195, but you transmit somewhere between 14.200 to 14.210 MHz!

Early on, outboard 2nd VFO's were offered as an option, but now almost all modern transceivers come equipped with two VFO's. So for split operation, set VFO (A) to the DX frequency and set VFO (B) to somewhere between 14.200 to 14.210 MHz.
Also the rig's RIT may cover the split, particularly on CW, as CW splits are usually only in the order of 2 to 3 kHz away from the DX frequency.

Many phone operations will require a greater range so the two-VFO configuration is used. Many rigs have a function button which allows you to momentarily listen on your choice of transmit frequencies This is TF Set on the Kenwoods. With TF Set, the Main tuning dial is varied to check for a likely transmit choice, and this does NOT change the receive VFO (A) frequency. Use your TF SET control to figure out if there is a pattern to how the DX station is picking out the callers. Maybe you want to avoid calling on a frequency where a Big Gun is parked, then again the Big Gun is likely to get through the pileup well before others and you might want to tail-end there or be up/down 2 or 3 kHz ready to call when the Big Gun works em.

For example you hear the DX pick out AC6V, quickly you hit TF SET and tune with the main dial to see if you can hear AC6V give his report. Repeat this a few times and determine if the DX is moving up, down, by how much or maybe staying right on the same frequency as the last caller. And curses, he may be just tuning randomly.

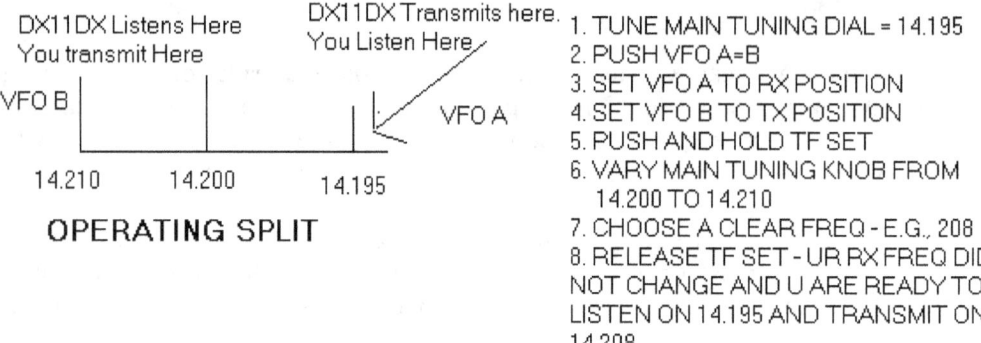

1. TUNE MAIN TUNING DIAL = 14.195
2. PUSH VFO A=B
3. SET VFO A TO RX POSITION
4. SET VFO B TO TX POSITION
5. PUSH AND HOLD TF SET
6. VARY MAIN TUNING KNOB FROM
   14.200 TO 14.210
7. CHOOSE A CLEAR FREQ - E.G., 208
8. RELEASE TF SET - UR RX FREQ DID
   NOT CHANGE AND U ARE READY TO
   LISTEN ON 14.195 AND TRANSMIT ON
   14.208.

Two strategies for working a split pileup are:

1) Pick a clear spot and holler away, sooner or later the DX station will tune into you, hopefully before they go QRT.

2) Determine the pattern that the DX station is using for working the pileup. The DX may be tuning up in 2 kHz increments, so you want to transmit up a bit from the last successful caller. Or tail end by being right on the responding stations frequency and throw in your call as a tail-ender. If the DX station is tuning at random, best pick a clear frequency and stay there. But – catch this: Some DX stations with a very large pileup say I'm listening "up 5" when he is really listening "up 10". This method divides the callers into two groups - the SMART ones who know how to LISTEN, and the other ones, who don't. Guess which group usually gets to work the DX?

For CW operations, the DX station is usually listening up 2 kHz or more and will identify a split CW operation by QRZ DX11DX UP, this will move the pileup away from the DX transmit frequency. The DX station will probably use RIT and tune up and down over a range of 2 to 5 kHz

The mob typically tunes up 2 kHz and sends their call a dozen times, so tune 2 to 5 kHz and determine any pattern the DX may be using to select callers, tail-ending works well in this situation.

5-9. THE BEST OPPORTUNITIES TO WORK DX

Peaked Propagation. Listen to DX signals and usually they will peak as a matter of propagation conditions and time. This can be short-term peaking – an hour or so, or may be the peak for an entire 24 hour period. For rare countries or DXpeditions, start a log as to when their signal is the strongest and schedule yourself to be in the pileup. This doesn't necessary mean when they are peaked at S-9. If they are inaudible for hours and typically peak to a perceptible level at a particular time, this can be an opportunity, particularly if the DX is working by district or so weak that few are calling. Optimum paths and superb propagation to you will beat folks elsewhere with the big amps and towers any day.

Working split. What if the DX picks up callers by frequency and not necessarily the strongest? Well that is exactly what happens in a split operation. Learn how to find a clear spot and how to track the DX pattern of tuning. See paragraph on split operations.

When your country is asleep. For the USA, working DX on a weekday at 2:30 in the morning can net DX that would have enormous pileups during normal "DX Working Hours". Hounding the night bands, 80M, 40M and 20M (when the SFI is high) can be happy hunting grounds. Even 10M can be open to midnight or longer when the SFI is approaching 300.

The Quiet Band. On occasion, the bands will be very quiet as far as atmospherics or person-made noise is concerned presenting an opportunity to work weak DX. Nearby electrical storms can be so severe as to make you want to turn off the radio, but at other times one can hear a marked reduction in atmospheric noise and it is amazing the weak DX one can hear. Person-made noise may abate during non-working hours or while it is raining or damp.

ESP (Extra-Sensory-Perception) contacts. Surprising how many DXers "can't hear the DX". Good headphones, DSP units, proper control settings, and special receiving antennas can just barely allow you to hear the "unhearable". When the DX signal starts to disappear down into the mud, the DX may call CQ with no takers and remember you are dealing with top-notch operators who are quite experienced in picking out weak signals. And on the islands they can hear quite well – no person-made noise or power line noise there. CALL!!

Operating habits. DX11DX comes on the air at the same time every night but Sunday at the same frequency, be there or be square. Check the DX Summit for habitual times of DX operators.

Poor propagation. If the DX has excellent propagation to an area of your country but not you (right or left coast), hang in there as the question "any West (East) Coast" may come your way.

Gray line, sunrise and sunset times are opportune times for DXing. See Chapter 4.

Tune Tune Tune. Don't rely too heavily on the DX Packet Clusters. One night here on the west coast, Turkey and Oman came in on 17M and not a single spot came on the cluster for 30 minutes after the DX appeared. TUNE TUNE – even "Dead Bands".

WARC Bands. For whatever reason, the 12M, 17M, and 30M bands appear to be used less than the popular 20M and 10M bands. A ZS on 10M or 20M will certainly cause a pileup, but on 12m they can be heard calling CQ with few takers and nice ragchews are in order here.

The No-Linear Band. On 30M everyone is equalized to 200 Watts, so no grousing is allowed about "Can't compete with the amps".

Read The DX Bulletins. DXers have other things to do in life, and the uninformed may not blast out their signals for several days after a DXpedition commences. The adage of "don't bother to call in the early days of a DXpedition" – is bad advice, get in there early.

Get in there late. HK0M is calling CQ CQ and no one answers! Why? Has he worked them all? Is every one working or asleep? No one can hear him? Well it is a common occurrence for a long term DXpedition to run out of takers in the last few days of the DXpedition.

Contests. Be sure to read the DX bulletins as to what rare DX will be on during a contest. Remember that they are on for many contest hours, so no hit and run hour operations here. Strange but ZC4 seems to be on during contests but seldom heard otherwise.

CW. Many DXpeditions or DX operators are on CW only. Learn the technique. Same for RTTY.

IOTA. Many DXers don't mess with IOTA contacts, but they count not only for IOTA, but also the parent country. TA0/IT9YRE Asia Iota # 154, Giresum Island counts for Turkey (I think).

Be Ready. The DX station changes bands – can you quickly follow them? Sometimes, forget tuning the linear – call first with 100 watts while others are tuning up. A packet cluster announcement appears for the first time, like a submarine, how fast can you "dive dive" and be one of the first there ?? Marking your linear tuning points can allow quick shifting of bands. Or sell your car and buy a no-tune linear!!

5-10. LOW BAND DXING

Low band experts will tell you there is a lot of DX to work on the low bands and it does not necessarily require big antennas and maximum power (although these will help). Both phone and CW can be worked and although these bands can be noisy, there are several techniques for receiving on these bands. There is a surprising amount of DX here, and like the higher bands; the secret is to know when to listen. During the hour of darkness the low bands are often open to various parts of the world depending on the time and season.

Antennas can be elaborate or simple vertical wires or tubing and with a good ground, good results can be achieved. By studying propagation characteristics and the operating times of the DX stations, you can reduce the amount of time you have to hound the bands for DX. But being primarily nighttime bands, you will have to give up some sleep late at night and early in the morning to take advantage of the propagation.

Contrary to popular belief, there can be lot of DX on the low bands during the summer. True that the low bands are noisier due to static and electrical storms, but there can still be excellent openings especially in the morning hours when the bands quiet down. The alignment of the gray line and the global darkness patterns are different for the seasons, so each season brings new DX opportunities. Using a computer program to predict propagation is a necessity and several are available on the Internet. Geoclock, Miniprop Plus, and the W6EL Propagation program are tools that show the Gray Line which is the period of semi-darkness that is created as the Earth rotates through day and night cycles. This Gray Line or "Terminator" changes with the seasons as the tilted axis Earth rotates around the Sun. For more on Gray-Line DXing – see Chapter 4.

The optimum time to work DX is when both ends of the path are in near darkness which occurs when they are both in their respective Gray Lines.

This includes long path openings which can occur in fall and early winter giving long path openings to Europe around our sunrise and their Sunset.

Contests and DXpeditions are excellent opportunities to work the low bands as DX stations are on the air at all hours and on all open bands. Contest operators are very savvy as to openings and know the sunset/sunrise times throughout the world. Checking the DX Packet Clusters and the DX Summit will allow you to determine who is on the air and at what times folks in your area are working them.

Low band DXing frequencies tend to be more specific than the higher bands. On CW, most DX stations frequent the lower end of the bands; usually the bottom 10 KHz. Contesters can be found up to 30 KHz from the lower edges. On single side band, DX windows are as follows:

40 Meters. Usually, 7050 to 7100 kHz but mostly nearer 7050 with the DX listening on their frequency and/or an announced split in the US phone band.

80 Meters. The DX window is 3790 to 3800 kHz and this is where most of the activity occurs. Many countries do not allow Amateur operation above 3800 KHz although some DX such as South America and some Pacific can be found above 3800 and down as low as 3775. This stretches out during contest times.

160 Meters. 1800 to 1850 kHz for both CW and SSB as many countries only allow amateur operation in this narrow frequency range.

Atmospheric and person-made noise can be a big problem on the low bands. Try using your noise blanker and adjust the tone and IF shift controls for optimum listening. External audio band-pass noise filters and noise cancellers are available. DSP units both internal and external can help as well. And of course, turn down the RF Gain control for optimum listening. If you have a horizontal high-band antenna, try using it for receive to improve the signal to noise ratio. Specialized receiving antennas such as loops and beverages are in order for optimum listening. These can be found in ON4UN's book "Low Band DXing". Also see AmCom's new noise canceller – ClearSpeech in World Radio magazine July 2001.

5-11. WORKING CONTESTS

Contacts made during a contest do count for awards. Send QSLs the same as you would for a DX contact including contest info such as the serial number. Contesters are good QSLers and DXCC entities such as Western Sahara, ZC4, and others quite often appear during a contest but otherwise may be infrequently heard.

You do not have to join a club or officially enter a contest to participate. You can participate in a contest whether or not you decide to formally submit an application and your logs. There are easily over 200 contests per year - four contests per week! Besides magazines and publications, one of the best sources of contest activity is LA9HW, Jan Almedal's Contest Calendar Pages at URL: http://home.online.no/~janalme/hammain.html   Includes rules.

Contesting is an art unto itself if you are going into the contest seriously as the callor. It is a grueling discipline requiring a big station, proficiency in logging by hand or with a program such as CT Contest. Many hours are needed to prepare for a contest – setting up and checking equipment, creating a quiet environment for computers and packet and getting ready for long hours of operating. Maintaining a high Q-rate (contacts per hour) is a gift or an art depending who you talk to.

Some contests offer awards for the smaller stations running 100 Watts or QRP. In a DX contest you are probably better off as a small station to be the callee rather than the callor. But in the WPX contests with a goofy prefix (AC6) you will get lots of action.

Since contesting can be very involved, best is to practice as the callee and then join a contest club to learn the art of big gun contesting. You will be amazed at the knowledge that the contesters have regarding propagation, best times to work various continents, achieving high Q-rates, computer logging, etc.

One of the easiest ways to work DX is in the contests. You don't have to be in the contest to participate. Some of the best operators in the world often travel to remote locations, plus there are the "super contest stations" in various countries operating with enormous antennas and state of the art equipment. They know how to dig out the weakest of signals, and are happy to get a few extra contest points. At the beginning of a contest, it may be tough getting thru the pileups, but often the big stations go begging for contacts during the last hours. Some very rare countries come on during contests but not otherwise heard often. Example SO2 Western Sahara, and ZC4.

Get some CW Contest pileup tapes – and practice the protocol. See Chapter 3, Operating Aids. With your computer have it send your call repeatedly gradually increasing speed up to 30 wpm + until your call is a musical sound - not dits and dahs and you can copy it at just about any speed. For CW Contests, you should be able to recognize your call at any speed – so you'll know when they come back to you. Just reply TU ur 5NN and the report (zone, etc.).

Or push buttons on your memory keyer. Be sure to listen if the DX is working split (up) and what continent or district is being called for. If not sure, check the DX Packet Clusters, folks often report these.

One thing for sure, the best way to learn how to work DX during a CW contest is to jump in. If you goof up, the contest station will call you perhaps twice and then move on if you don't reply. The DX op doesn't know if you had a power or equipment failure, landline, or whatever – and they really don't care, so don't be intimidated or nervous. One of the holdbacks seems to be a newbie will screw up nervously in sending the high-speed code required.

Get a memory keyer, you won't believe how many hams use these all the time. It is really is a "push a da button" deal once you get the hang of it. There is very little chance of the DX pausing to ask a question, so don't worry about it, just send TU 73 and disappear.

Note that some big contest stations operate just before the contest as a warm up – great time to get a contact. Likewise after a contest.

Contests are not allowed on the WARC bands, but many big contesters will operate the WARC bands before/after the contests to give others a new one.

Be the callor instead of the callee. A fun contest where you can be the "DX Station", the Callor as it were, is the CQ Magazine WPX contest. With a goofy semi-rare prefix like AC6, NG7, etc., the DX will call you because your prefix counts as multipliers. Not that they won't call a K6 or W4 as they count for Q's also but the semi-rare prefixes will cause pileups. Have had DX like Easter Island and Kenya call me during the WPX contest.

If you are a fairly good CW operator, the WPX CW Contest can be lots of fun as the Callor. Keep the bandpass open a little to hear the folks calling off frequency, and use your RIT control to work them. To find a clearer frequency to operate on - use your XIT to QRL it, then move to the clear frequency. This is not to suggest you can't be the callor in a DX contest, but it usually takes a big station and the USA is not exactly DX, best saddle up and go to the Caribbean for this joy.

It is very common to send RST reports in abbreviated form during a contest. These are called cut numbers or abbreviated numbers. For example 599, is sent as 5NN. "N" in place of the number "9". Also another time saver is for the zero using a long "T". "T" is sent in place of the number zero as in " POWER HR IS 3TT WATTS".There is a number code for all numbers; however, the N and T codes are the most common ones. Also CW stations sometimes report their zones as "A4" or "A5" instead of sending "14" or "15". 1 = A, 2 = U, 3 = V, 4 = 4, 5 = E, 6 = 6, 7 = B, 8 = D, 9 = N, 0 = T

There is much more advice for working DX during contests in Chapter 7 and 8.

5-12. QRP OPERATION

Why the fascination with operating with low power. After all "isn't QRP for sissies" or "I didn't climb to the top of the DX food chain to eat lettuce". Well it is a challenge you might not experience with a full gallon. It requires extreme patience and good DX ears and it is very good practice for weak signal DX work. QRP is defined loosely as with a power output of 5 watts or less (10 watts peak power for SSB). QRPp is defined as accurately measured to be less than 1W.

When the propagation gods are with you, working Africa with 5 Watts of power is achievable on the upper bands, getting progressively more difficult on Kilowatt Alley (20M) and below in frequency. It is very satisfying to work 100 countries QRP, much more demanding than DXCC with 100Watts.

There are several QRP contests and some of the big contests have QRP entries. Many awards are available for working 100 countries QRP.

Also when you are down to chasing the top 100 DXCC entities or so and nothing new is on at the moment, maybe try QRP. Modern rigs can be turned down to 5 Watts, or it can be fun to get back into home-brewing a QRP rig like ye olden days of building your own equipment.

The QRPp Low Power Long Distance Record is held by KL7YU and W7BVV using one MicroWatt over a 1,650 mile Ten Meter path between Alaska and Oregon in 1970. This is the equivalent of 1.6 BILLION Miles per Watt!! A QRPp award is available for those that work over a distance of 63 miles (100 kilometers) between the applicant's transmitting antenna and the receiving station.

Contest tip on QRP. In a non-QRP contest, do not call using your callsign/QRP. Contesters don't want to send the QRP part or mess with acknowledging your QRP operation. Just put the /QRP on your QSL card.

QRP calling frequencies and nets are available at URL: http://www.qrparci.org/

5-13. THE CONTROVERSIAL DX NETS – DX BY MENU

A fast way to get the basic DXCC certificate (100 verified contacts) is to make use of the DX nets that are very popular across the HF bands. Hundreds gather here each day to participate. Here a net control will periodically call for check-ins of DX stations, then they ask for those desiring to work the DX. From here a list is created to contact the DX countries one operator at a time. Nets are well suited for those with low power stations since it gives each participant a clear frequency to attempt their contacts. Two popular nets are found on 14.247 MHz and 21.355 MHz. More DX HF Nets at URL: http://ac6v.com/nets.htm

Many DXers are very critical of the DX nets, but have been known to jump in if the DX is extremely rare. In the past, some countries with just a very few operators, only came on nets being unable or not willing to run a pileup. So nets have their place and after all, do we "Real DXers" want 7,000 netters in our pileups? At best they are assisted contacts, but are perfectly legitimate and are recognized contacts by the ARRL DXCC Desk. But true blue DXers refer to it as spoon-feeding and shooting fish in the old barrel. Some claim the DX Packet Clusters are spoon feeding as well, but you still have to break the pileups, no nickels nickels here.

Be aware that your peers may look askance if you are a netter, and I wouldn't mention it at your first DX Club meeting. Early on in your DX career, you need to decide to remain pure and not work DX on nets or say hey it's a hobby and do your thing.

5-14. THE DON'TS OF PILEUPS

Although all is fair in love, war, and DX, observing the following list of don'ts will allow the DX station to realize a faster Q-Rate and hopefully work you sooner.

In split operations -- DO NOT transmit on the DX frequency asking where is he listening – tune around (usually up) and you'll know.

To be sure, just listen to the DX operator, they will announce it periodically e.g., this is DX11DX listening up 200 to 250, QRZ. Checking the Packet Cluster will also get this information.

DO NOT call if you can't hear the DX station. Sounds dumb but it is surprising how many folks do this, maybe hoping someone will help them "hear" the DX, e.g., L0GUN he is calling you.

DO NOT try and assist someone else with a callsign or RST exchange. If the DX is having trouble copying a station and they say, "again?" avoid the temptation to assist.

DON'T CALL during other people's contacts, at best it slows down the Q-Rate, may get you on the "Black List", besides being downright rude.

When you discover a rare DX or medium rare DX station ragchewing, don't insert a break unless the DX is asking for specific information that you can offer. There are probably 100+ others who would like to break in.

For breaking into a domestic QSO or into a "common DX" QSO, use common courtesy, if the QSO is between two old friends talking more or less about personal items, breaking in is usually not welcome. If the QSO is interesting to you and you have something to offer, breaking in is common enough, just break and when acknowledged, ask if you can join the QSO. Often a question is posed that neither have the answer for, and a "break info" would be in order.

When the DX is calling by district, DON'T call out of turn; you slow things down for EVERYONE. A sharp DX station will ignore you anyway. The DX should make it clear what he/she is asking for. Calling for the 4th call area invites any one with a W4 call or anyone that is portable W4 e.g., AC6V/4. Calling for "call letters W4" is not inviting the portables in the W4 area. Gets confusing but you decide.

DO NOT slow down the DX run by asking for QSL info – most DX operators give it periodically – if the DX runs forever without giving QSL info and you haven't seen it on the cluster or QSL routes on the web, then ask. On the packet clusters, you can query by typing in SH/QSL DX11DX and get the info. The DX Summit is good for this too.

Some DXers recommend that you should be able to understand CW well enough to copy something besides your own call and 599. If you can't UNDERSTAND what the station is sending, even when hearing them well, it's time for some practice. You should be able to copy what district or country is being worked. Editor Note: Not so sure about this as many CW pileups are your own call and 599 and that's it. However, it should be clear what district or country is being worked, check the DX Packet Cluster – someone will probably announce when the DX is calling your district and country, just be sure it is current packet announcement.

Stay in sync with the rhythm of the pileups, give the DX a chance to answer and others can hear.

DON'T TUNE up on the DX frequency or the split frequency. DO NOT over adjust your transmitter and spatter all over the band.

DO NOT ask, "What is his call". In this age of DX Packet Clusters it is incredible any one would ask. Good DX operators give it out frequently – just listen. Here again check the clusters.

DON'T play Radio Cop (police persons) – typically one poor guy is out of line and 10 cops are telling him everything from good advice to insults. Best not to do this as your rock-crushing signal covers up the DX and creates chaos and just slows things down.

During a pileup, don't ask the DX to QSY to another band for a new band country, a coupla hundred Dxers will hate you for this.

When rag-chewing with DX, avoid talking about politics, cultural differences and religion.

5-15. DX PACKET CLUSTERS, TELNET AND THE DX SUMMIT

Before repeaters, DX brethren would have a network on VHF simplex to inform each other of DX. This was supplemented with the infamous "one-ringer" – a phone call at 2:30 in the morning alerting the conked out DXer to get on the air. Buddies kept lists of each other's needs and the "one-ringer" system aggravated many an XYL. In the 70's exclusive DX 2M repeaters came into being, all QSO's were more or less DX oriented and DX spots took priority.

In the 80's, a new phenomenon came in to being – the DX Packet Cluster. Using computers and VHF packet radio with AX.25, one station is set up with software and is linked to one or more other stations who have installed the same system. These nodes when connected are called a cluster. DX announcements propagate thru the cluster to all users. Later the clusters interconnected to the next outlying one and soon statewide DX Packet Clusters were in operation. Samples of DX Packet Cluster announcements today are:

```
10113.9  D68C          13-Feb-2001 2257Z  up
28020.9  VP5/N2GA      13-Feb-2001 2244Z  QSX 28021.40 george sm up
28021.8  D68C          13-Feb-2001 2241Z  QSX up ONE !
10106.2  K2LS          13-Feb-2001 2238Z
7003.8   D68C          13-Feb-2001 2249Z  wrkd 5 Watts
```

Some Packet Cluster abbreviations:
CBA – Call Book Address
QSX xxxx - The split frequency where the DX is listening
EU Only – DX is listening for Europe only
NA (SA) Only - DX is listening for North America (South America) Only
NIL – No Signal Heard
UP – Split Operation - DX is listening UP (usually 5 to 10 kHz SSB, 2 to 5 kHz CW)
MNG (MGR) – QSL Manager of the DX Station
VIA - QSL Manager of the DX Station
DIRECT – DX is listening and transmitting on the same frequency
SIMPLEX - DX is listening and transmitting on the same frequency
CLG (CQ) – DX is calling CQ
HR = here
WKD (WRKD) – Worked (A Brag Spot)
(NA) (SA) (AS) (OC) (EU) (AF) (AN) continents - North America, South America, Asia, Oceania, Europe, Africa, Antarctica
AF-032 - IOTA Number for continents e.g., AF-032 is the Islands On The Air Number for Zanzibar Island off the African Coast
A "busted" spot is one in which the callsign sent to the cluster is wrong - not the callsign the DX station is actually using. Was a typo or the spotter miscopied it.
WARC – The WARC Bands are 30 meters, 17 meters, and 12 meters   /B - Beacon

Ah some goodies there, but with the DX Packet Clusters, the pileups build up very quickly, best you find them, work them, then spot the DX on the cluster. A beginner may be tempted to spot them first out of excitement, but that's like being in a football game and telling the defense where you are going to throw a pass. For the DX Packet Cluster frequencies in your area - contact the local DX club or a Local DXer. You will need a computer, a TNC or soundcard and software, and a 2M transceiver for on-the air DX Packet Cluster reception.

Later all of this information was available on the Internet under the media of telnet. Telnet is the Internet standard for remote login service. Just configure your browser for telnet operation, see the help menu on Internet Explorer or Netscape. A massive listing of DX Telnet connections for just about every state is available at VE9DX's page – URL:
http://ve9dx.weblink.nbtel.net/telnet/sites.html

With today's DX Packet Cluster, one can send and receive DX spots and query the node for QSL information, MUF, distance, bearing, periodic propagation reports, and a whole host of features including a bulletin board where one can announce the latest QSL card received or list equipment for sale. The system allows personal talk messages, send and receive mail messages, search and retrieve archived data, and access data from information databases. Information on log-ons, protocol and commands are discussed in detail at the help menus.

In addition a website featuring an on-line international DX spotting network (The DX Summit) came into being several years back, OH2BN now OH2AQ. Even if you don't have propagation, you can determine if your desperately needed country is even on the air and get an idea of their operating habits and frequencies. See DX Summit at URL: http://oh2aq.kolumbus.com/dxs/
Some samples are:

14195.0	3Y0C	in aprox 10 mins	2308 13 Feb
28495.0	V73ZZ		2308 13 Feb
28462.1	C6AGS	via ki6t	2308 13 Feb
3799.0	LU1IV	fb sigs	2310 13 Feb
3788.5	SO0DIG		2309 13 Feb
1822.0	D68C	with 100 watts	2306 13 Feb

On the DX Summit, one can query The Spot Database Search to determine when a particular entity is on the air. Example search for EK will show all Armenia spots for several weeks.

Although old timers grouse about "it ain't like it used to be", the packet clusters are invaluable for the budding DXer. Old timers are more likely to announce the rare ones, but the new guys even announce Japan. Thousands of DX spots occur during a contest. Without the DX Packet Clusters, you are kinda at the bottom of the DX chain. But don't get too attached to the clusters, lots of DX can be found by tuning before it hits the clusters. When using the cluster spots, make sure you have the DX call correct. It is easy to see a spot for a rare DX station, jump on the frequency, beat the pileup, and log the call wrong because the call posted on the cluster has a typo or just plain wrong. Listen well after the contact to double check the DX callsign

5-16. DX CENTURY CLUB

The DX Century Club (DXCC is copyright and registered by the ARRL) is the premier operating award in all of Amateur Radio. In 1937, the ARRL introduced the DXCC Program, which was discontinued during WWII and started all over again after the war. The basic certificate (which can be obtained in several categories -- mixed modes, phone, CW, RTTY, Satellite, 160, 80, 40, 20, 15, 10, 6 and 2 meters) is awarded for working and confirming at least 100 entities on the ARRL DXCC List. Be aware that all HF band QSO's including 6 meter and 2 meter QSO's can be submitted for credit. So WARC QSO's do count. Endorsements are available in specific increments beyond the 100-entity level, culminating in the coveted DXCC Honor Roll, for those at or near working them all! Further DX Century Club information is available at http://www.arrl.org/awards/dxcc/. The top most wanted DXCC countries are listed in Appendix A-12. The more common DXCC entities are also listed in Appendix A-12.

For those who enjoy the thrill of the hunt on more than one band, the 5-Band DXCC can be earned for working 100 entities on 80, 40, 20, 15 and 10-meter bands.

5BDXCC is a good test of the DXer's operating abilities, but is well within the reach of all of those willing to work for it. 5BDXCC qualifiers receive a handsome certificate and are eligible for a very attractive plaque. The 5BDXCC award is endorsable for the 160, 17, 12, 6 and 2-meter bands.

5-17. WORKED ALL STATES

Work and confirm all 50 states in the USA in any combination of bands/modes will earn the basic certificate. Specialty certificates are issued for a variety of different bands and modes such as Satellite, 160 meters, SSTV, RTTY, each VHF band. Available endorsements include SSB, CW, Novice, QRP, Packet, EME, and any single band except 30 meters. To encourage increased activity and station improvement throughout the bands, the 5-Band WAS certificate (and plaque) is available for working all states on 5 amateur bands (except 12/18/24 MHz).

5-18. WORKED ALL CONTINENTS

Sponsored by the International Amateur Radio Union (IARU), the Worked All Continents award is issued for working and confirming all six continents (North America, South America, Oceania, Asia, Europe and Africa) on a variety of different bands and modes. A 5-Band WAC certificate and a 6-Band sticker are also available. See URL: http://www.arrl.org/awards/

5-19. CQ MAGAZINE AWARDS

CQ Magazine Awards include DX Awards, Worked All Zones Awards (WAZ), Worked All Prefix (WPX) Award, County Hunters (USA-CA Award), and the CQ Millennium Award.

WAZ is a prestigious award in so much as it shows you have worked and confirmed "The Four Corners of the World" (all 40 Zones). See Appendix A6 for a map of the 40 CQ Zones. For details on the CQ awards, see URL: http://www.cq-amateur-radio.com/infoc.html

5-20. WORLD RADIO MAGAZINE AWARDS

The World Radio Magazine sponsors several awards including a PSK31 award and the 100 Nations award. See URL: http://www.wr6wr.com/

5-21. ISLANDS ON THE AIR (IOTA)

The IOTA programme is sponsored and administered by the Radio Society Of Great Britain (RSGB). This is another form of DX chasing and there is a lot of IOTA activity on the bands. Look for IOTA spots on the DX packet clusters or the DX Summit. Also there are defacto IOTA operating frequencies and nets as listed below. Although there are thousands and thousands of islands throughout the world, the IOTA programme is limited to 1200 groups. For more information see URL: http://www.rsgbiota.org/index.php4

IOTA CW: 3.530, 7.030, 10.115, 14.040, 18.098 and 21.040 MHz
IOTA SSB: 3.755, 7.060, 14.260, 18.128, 21.260, 24.950, 28.460 and 28.560 MHz
US Island Hunters: 7.250, 14.250 to 14.260 (main), 21.350, 28.450 MHz

5-22. AWARDS AND WALLPAPER

Ten-Ten International has several awards, see URL: http://listserv.lehigh.edu/lists/tenten-l/
Most countries have awards and wallpaper and the definitive source is K1BV's Award Directory at URL: http://www.dxawards.com/

5-23. DX NEWS LETTERS AND MAGAZINES

Before the advent of the Internet, DXers subscribed to printed DX newsletters which were of necessity rushed by airmail to supply the latest news. There are still several of these available in printed form. But today, many are available free on the Internet.
By reading these periodically, you will always be aware of the latest DXpeditions and activity of rare DX stations. Frequently the operating frequencies and times are given as well as QSL information. Newsletter URL's are at http://ac6v.com/newsletters.htm#DXN These include:

Announced DX Operations By Bill Feidt/NG3K
DX Web Site - From HFRADIO.ORG - Has DX news and more
The Ohio/Penn DX Bulletin. Internet Edition by KB8NW
The 425 DX News Home Page. - Edited by I1JQJ & IK1GPG
The 425 DX News Bulletins & Archives.
IOTA DXPEDITION NEWS - By Chris N1HRW
The ARRL DX Bulletin. - W1AW DX Bulletins
DXBands.com News - Receive weekly e-mail newsletter
DX Central - Latest Breaking DX News
RSGB News - DX, Contest, Propagation News

5-24. PREFIXES

As you tune the bands and hear the alphabet soup of the DX calls or see them on the DX Packet Cluster, you need to know the DX country prefixes or at least have a handy chart. This helps with your beam headings and will determine if you even want to work them. For example, you hear a ZZ9DX (probably during a contest), and after consulting your list you find it is Brazil, normally a PY#xx.

So be aware there are "primary" prefixes and perhaps several secondary prefixes as well. Examples; Indonesia is usually YB but could be from the following ITU block allotments (7AA-7IZ, 8A-8I, JZ, PK- PO, YB-YH). These are typically allotted during contests or special events (900th anniversary of the revolution!!) Appendix A10 contains all known prefixes. As you progress along with DXing, you might want to apply for the CQ WPX Award, so working common and oddball prefixes might be worth your while.

5-25. RARE, MEDIUM RARE AND WELL DONE DX

When one is just starting into DXing, the rare ones are the ones you haven't worked or confirmed – all of them. But after a few months, patterns will develop where you will find a lot of common (well-done DX) such as England, Germany, France, Belgium, Brazil, Argentina, Japan, Canada and so on. So what is rare, medium rare and well done in the way of DX?

Appendix A12 shows the countries with Ham populations over 1,000 and thus considered commonly heard on the bands. The rarest 100 country entities are shown in Appendix A12. Others might be considered medium rare DX. Many of the rare DXCC entities are places that are uninhabited and require DXpeditions to place the entity on the air.

Others are countries with very low Ham populations and/or with Communication Authorities that grant very few, if any, operating permits.

When a rare one shows up, you should endeavor to work them at all costs as it could be months or years before another opportunity occurs.

A good example is Navassa Island, NP1, which is administered by the Fish and Wildlife Service, US Department of the Interior. Permits to operate there are difficult to come by and it has been several years since the last operation. Desecheo, and Howland-Baker are others that are difficult to obtain permission for a DXpedition. Another example is the last accepted DX operation from VU7 Lakshadweep Islands around 1989 to 1990. The author worked a TT8 in the early 80's and neglected to send for a card immediately, it was 20 years later before another TT8 was contacted!! Get 'em while they are hot!

Slims, Pirates, and Bootleggers.   Over the years there have been some amazing illegal operators who were so adept at CW or voice as to fool a good many DXers. These are the hackers of the DX airways.

Slim - someone pretending to be a rare DX station. For example, someone in southern Argentina pretending to be on Heard Island, VK0IR

Bootlegger - usually not a Ham but a wanna-be, making up a call sign and getting on the air, usually not in the call book, Sometimes it is someone who already bought a radio, took the test and flunked, and then gets on the air anyway.

Pirate - Someone using an existing call sign and operating on the air, e.g. claiming to be WA6YOO/4 on a North Carolina island IOTA group. The real WA6YOO is suddenly overwhelmed with QSL requests. These definitions are somewhat flexible because some of their aspects might be combined. Experienced DXers can spot an illegal operation by means of reported bearings, signal strength, propagation patterns, and triangulation amongst several listening stations. Be leery of slim and pirate reports on the DX Packet Clusters, could be real.

5-26. SLASHED CALLS

SLASH CALLS. AC6V/4 means I have a California call (6), but I'm operating portable in the USA fourth call area or now live in the 4th call area permanently. Not required under FCC rules that I identify in this way, however. In DXing, it is best to use /# if you are operating out of your callsign area as the DX may call by district. Europeans sometimes use P8, e.g. DJ#xxx/P8 meaning they are portable in their 8th call district and not North Korea!!

An English Ham operating in the Bahamas might sign G#xxx/C6A or C6A/G#xxx, I have seen it both ways. Each country may have different rules. And one can be portable in their own call area AC6V/6 means I'm away from home operating somewhere in California (maybe a rare county).

For the CEPT agreement it states, "When transmitting in the visited country the license holder must use his national call sign preceded by the CEPT call sign prefix. The CEPT call sign prefix and the national call sign must be separated by the character "/" (telegraphy) or the word "stroke" (telephony). For a mobile Amateur Radio station, the national call sign must be followed by the characters "/M" (telegraphy) or the word "mobile" (telephony).

For a portable amateur radio station, the national call sign must be followed by the characters "/P" (telegraphy) or the word "portable" (telephony)". For operating abroad, see URL http://www.arrl.org/FandES/field/regulations/io/#cept /MM is for Maritime Mobile and these do not count for DXCC.

5-27. ZONES (See Appendix A6 and A7)

For Amateur Radio, the world is divided up into zones. There are two major zone assignments that are of interest to DXers, the CQ zones and the ITU zones. For the CQ zones there are 40 zones and for the ITU zones there are over 90 zones. For certain contests, you will need to know your zone number as part of your report. For example, W6's are in CQ Zone 3, but in ITU Zone 6. A contest may require one or the other. CQ zones are CQ Magazines zone delineation, while the International Telecommunication Union is an official body.

For example, for the CQ World Wide DX Contest your report is RS(T) report plus CQ zone. For the YO DX Contest, the Exchange is signal report and ITU Zone.

Also some contests, particularly VHF contests, may require grid squares. Your grid square can be found at QRZ.com. Mine is DM13IF. Also see URL: http://www.icomamerica.com/ Click on AMATEUR, then U.S. Grid Square.

In addition, if you intend to "be in the contest" you will want to know what zone the DX station is in, for your multipliers.

Later you may want to apply for the Worked All Zones Award (WAZ) sponsored by CQ magazine, so keep track of the zones you work in your log book and you might as well send out Bureau cards for these – its cheap enough via the burro. Working all zones on a single band can be tough - tougher than DXCC for sure. Maps of CQ zones and the ITU zones are on the web, also see Appendix A6 and A7. Also the ARRL offers the "Radio Amateurs World Atlas" with all DXCC entities, but only shows CQ zones. Excellent and highly recommended.

5-28. USA BAND PLANS AND POWER LIMITATIONS

For the USA, the FCC sets the frequency allocations, modes allowed, and the power limitations for USA Amateurs. For other countries, see your Communication Authority Rules and Regulations.

POWER LIMITATIONS USA

At all times, transmitter power must be the minimum necessary to carry out the desired communications. Unless otherwise noted, the maximum power output is 1500 watts PEP. All classes are limited to 200 watts PEP in the 80, 40, and 15 meter Novice/Technician Plus subbands. Geographical power restrictions apply to the 70 cm, 33 cm and 23 cm bands; see *The FCC Rule Book* for details.

5-29. HF Thru 6M Frequency Allocations For USA Amateurs
(See FCC Part 97 for VHF and up) - Effective 15 Dec 2006

160 Meters
General, Advanced, Amateur Extra
licensees:
1.800-2.000 MHz: CW, Phone, Image,
RTTY/Data

80 Meters
Novice and Technician Plus classes:
3.525-3.600 MHz: CW Only
General class:
3.525-3.600 MHz: CW, RTTY/Data
3.800-4.000 MHz: CW, Phone, Image
Advanced class:
3.525-3.600 MHz: CW, RTTY/Data
3.700-4.000 MHz: CW, Phone, Image
Amateur Extra class:
3.500-3.600 MHz: CW, RTTY/Data
3.600-4.000 MHz: CW, Phone, Image

60 Meters

The FCC has granted hams secondary access on USB only to five discrete 2.8-kHz-wide channels. Amateurs can not cause inference to and must accept interference from the Primary Government users. The NTIA says that hams planning to operate on 60 meters "must assure that their signal is transmitted on the channel center frequency." This means that amateurs should set their carrier frequency 1.5 kHz lower than the channel center frequency.

General, Advanced and Amateur Extra classes:

Channel Center	Amateur Tuning Frequency
5332 kHz	5330.5 kHz
5348 kHz	5346.5 kHz
5368 kHz	5366.5 kHz
5373 kHz	5371.5 kHz

5405 kHz (common 5403.5 kHz US/UK)

USA Amateurs may use USB *only* with a maximum effective radiated power (ERP) of 50 W. Radiated power must not exceed the equivalent of 50 W PEP transmitter output power into an antenna with a gain of 0 dBd.

UK report of 200 watts (23dBW) PEP Using Morse, Telephony, RTTY, Data, Fax and SSTV.

40 Meters

Novice and Technician Plus classes:
7.025-7.125 MHz: CW Only
General class:
7.025-7.125 MHz: CW, RTTY/Data
7.175-7.300 MHz: CW, Phone, Image
Advanced class:
7.025-7.125 MHz: CW, RTTY/Data
7.125-7.300 MHz: CW, Phone, Image
Amateur Extra class:
7.000-7.125 MHz: CW, RTTY/Data
7.125-7.300 MHz: CW, Phone, Image

Note: Phone and Image modes are permitted between 7.075 and 7.100 MHz for FCC licensed stations in ITU Regions 1 and 3 and by FCC licensed stations in ITU Region 2 West of 130 degrees West longitude or south of 20 degrees North latitude. See Section 97.307(f)(11). Novice and Technician Plus licensees outside ITU Region 2 may use CW only between 7.050 and 7.075 MHz. See Section 97.301(e). These exemptions do not apply to stations in the continental US.

30 Meters

Maximum power, 200 watts PEP. Amateurs must avoid interference to the fixed service outside the US.

General, Advanced, Amateur Extra classes:
10.100-10.150 MHz: CW, RTTY/Data

20 Meters

General class:
14.025-14.150 MHz: CW, RTTY/Data
14.225-14.350 MHz: CW, Phone, Image
Advanced class:
14.025-14.150 MHz: CW, RTTY/Data
14.175-14.350 MHz: CW, Phone, Image
Amateur Extra class:
14.000-14.150 MHz: CW, RTTY/Data
14.150-14.350 MHz: CW, Phone, Image

17 Meters
General, Advanced, Amateur Extra classes:
18.068-18.110 MHz: CW, RTTY/Data
18.110-18.168 MHz: CW, Phone, Image

15 Meters
Novice and Technician Plus classes:
21.025-21.200 MHz: CW Only
General class:
21.025-21.200 MHz: CW, RTTY/Data
21.275-21.450 MHz: CW, Phone, Image
Advanced class:
21.025-21.200 MHz: CW, RTTY/Data
21.225-21.450 MHz: CW, Phone, Image
Amateur Extra class:
21.000-21.200 MHz: CW, RTTY/Data
21.200-21.450 MHz: CW, Phone, Image

12 Meters
General, Advanced, Amateur Extra classes:
24.890-24.930 MHz: CW, RTTY/Data
24.930-24.990 MHz: CW, Phone, Image

10 Meters
Novice and Technician Plus classes:
28.000-28.300 MHz: CW, RTTY/Data--
Maximum power 200 watts PEP
28.300-28.500 MHz: CW, Phone--Maximum
power 200 watts PEP
General, Advanced, Amateur Extra classes:
28.000-28.300 MHz: CW, RTTY/Data
28.300-29.700 MHz: CW, Phone, Image

6 Meters
All Amateurs except Novices:
50.0-50.1 MHz: CW Only
50.1-54.0 MHz: CW, Phone, Image, MCW,
RTTY/Data

A nice color graph chart of USA Amateur
Frequencies can be found at URL:

www.arrl.org/FandES/field/regulations/Ham
bands_color.pdf

5-30. CLASSIFICATIONS OF EMISSIONS

Classification of emissions -- From The FCC -- The Code of Federal Regulations, Title 47, Volume 1, Parts 0 to 19

Radio emissions are defined with an alphanumeric code, for example:

A1A is an ON/OFF keyed carrier (as in "CW" "Morse")
A3E is double-sideband with full carrier ("AM", "broadcast")
J3E is single-sideband ("SSB") suppressed carrier
F3E is frequency modulation telephony
F2D frequency shift keyed audio tone, data packet. Packet Radio.

COMMONLY USED TYPES ARE:

A1A -telegraphy (on - off keying) without modulation by an audio frequency;
A1B - amplitude modulation telegraphy with automatic reception, without using a modulating subcarrier;
A1D -amplitude modulation data transmission, double sideband, without using a modulating subcarrier;
A2A -on-off keying telegraphy using one or several modulating audio frequencies, or on-off keying telegraphy of an amplitude modulated emission;
A2B -amplitude modulation telegraphy with automatic reception and using on-off keying of the modulating subcarrier;
A2D -amplitude modulation data transmission, double sideband and using a modulating subcarrier;
A3C -amplitude modulation facsimile; the main carrier is modulated either directly or by a frequency modulated subcarrier;
A3E -amplitude modulation telephony, double sideband;

C3F -television, in amplitude modulation, with vestigial sideband;

F1A - telegraphy (keyed by frequency variation);
F1B -frequency modulation telegraphy with automatic reception, without using a modulating subcarrier;
F1D -frequency modulation data transmission, double sideband, without using a modulating subcarrier;
F2A - on-off keying telegraphy of an audio frequency for frequency modulation, or by on-off keying of a frequency modulated emission (special case: unkeyed frequency modulated emission);
F2B -frequency modulation telegraphy with automatic reception and using on-off keying of the modulating subcarrier;
F2D -frequency modulation data transmission, using a modulating subcarrier;
F3C - modulation frequency facsimile, by direct modulation of carrier frequency;

F3E - frequency modulation telephony;
F3F - television with frequency modulation;

G1D -phase modulation data transmission, without using a modulating subcarrier;
G2D -phase modulation data transmission, using a modulating subcarrier;
G3C -phase modulation facsimile:
G3E -phase modulation telephony;
G3F -television in phase modulation;

J1D -amplitude modulation data transmission, single sideband, suppressed carrier, using a modulating subcarrier;
J2A -on-off keying telegraphy using one or several modulating audio frequencies, or an on-off keying telegraphy, single sideband, with suppressed carrier;
J2D -amplitude modulation data transmission, single sideband, suppressed carrier without using a modulating subcarrier;
J3C - amplitude modulation facsimile; single sideband, with suppressed carrier;
J3E -amplitude modulation telephony, single sideband, with suppressed carrier;
J8E -amplitude modulation telephony, independent sidebands;

K1A -on-off keying telegraphy of a carrier transmitted by pulses, without modulation by an audio frequency;
K2A -on-off keying telegraphy of one or several audio modulating frequencies, or by on-off keying of a modulated carrier transmitted by pulses (special case: unkeyed modulated carrier transmitted by pulses);
K3E - pulse modulation telephony.

R3C - amplitude modulation facsimile; single sideband, with vestigial carrier;
R3D -amplitude modulation data transmission, single sideband, with vestigial carrier;
R3E -amplitude modulation telephony, single sideband, with vestigial carrier

5-31. CONSIDERATE OPERATOR'S FREQUENCY GUIDE

Upper and Lower Sideband Protocol

Except for 60 Meters (USB only), 160, 80, and 40 meters are traditionally lower sideband. From 20 meters up, it is upper sideband. Except for 60M, you can use either upper or lower sideband on any band where SSB is allowed. But not advisable as all signals are one or the other except yours. The protocol was a result of early SSB transceiver design as a matter of simplifying design

The following frequencies are generally recognized for certain modes or activities (all frequencies are in MHz). Nothing in the rules recognizes a net's, group's or any individual's special privilege to any specific frequency. Section 97.101(b) of the Rules states that "Each station licensee and each control operator must cooperate in selecting transmitting channels and in making the most effective use of the amateur service frequencies. No frequency will be assigned for the exclusive use of any station." No one "owns" a frequency. It's good practice and plain old common sense for any operator, regardless of mode, to check to see if the frequency is in use prior to engaging operating. If you are there first, other operators should make an effort to protect you from interference to the extent possible, given that 100% interference-free operation is an unrealistic expectation in today's congested bands.

A guide to where on the HF bands various modes and activities are generally found. All frequencies are in MHz.

Frequencies Modes/Activities

1.800-2.000 CW
1.800-1.810 Digital
1.810 QRP CW calling frequency
1.843-2.000 SSB, SSTV and other wideband modes
1.910 SSB QRP
1.995-2.000 Experimental
1.999-2.000 Beacons
3.500-3.510 CW DX window
3.560 QRP CW calling frequency
3.570-3.600 RTTY/Data
3.585-3.600 Automatically controlled data stations
3.590 RTTY/Data DX
3.790-3.800 DX window
3.845 SSTV
3.885 AM calling frequency
3.985 QRP SSB calling frequency
7.030 QRP CW calling frequency
7.040 RTTY/Data DX
7.080-7.125 RTTY/Data
7.100-1.105 Automatically controlled data stations
7.171 SSTV
7.285 QRP SSB calling frequency
7.290 AM calling frequency
10.130-10.140 RTTY/Data
10.140-10.150 Automatically controlled data stations
14.060 QRP SSB calling frequency
14.070-14.095 RTTY/Data
14.095-14.0995 Automatically controlled data stations
14.100 IBP/NCDXF beacons
Frequencies Modes/Activities

14.1005-14.112 Automatically controlled data stations
14.230 SSTV
14.285 QRP SSB calling frequency
14.286 AM calling frequency
18.100-18.105 RTTY /Data
18.105-18.110 Automatically controlled data stations
18.110 IBP/NCDXF beacons
21.060 QRP CW calling frequency
21.070-21.110 RTTY/Data
21.090-21.100 Automatically controlled data stations
21.150 IBP/NCDXF beacons
21.340 SSTV
21.385 QRP SSB calling frequency
24.920-24.925 RTTY/Data
24.925-24.930 Automatically controlled data stations
24.930 IBP/NCDXF beacons
28.060 QRP CW calling frequency
28.070-28.120 RTTY/Data
28.120-28.189 Automatically controlled data stations
28.190-28.225 Beacons
28.200 IBP/NCDXF beacons
28.385 QRP SSB calling frequency
28.680 SSTV
29.000-29.200 AM
29.300-29.510 Satellite downlinks
29.520-29.580 Repeater inputs
29.600 FM simplex
29.620-29.680 Repeater outputs

5-32. CALLING FREQUENCIES

160 METERS
1.830-1.840 CW, RTTY and other narrowband modes, intercontinental QSOs
1.810 QRP Calling frequency
1.840-1.850 CW, SSB, SSTV and other wideband modes, intercontinental QSOs
1828.5 -- DXpeditions CW Operations are frequently here

80/75 METERS
3.500-3.510 CW DX Window
3.505 DXpeditions CW are frequently here
3.560 QRP Calling frequency
3.590 RTTY DX
3.790-3.800 SSB DX Window
3.71 QRP Novice/Tech CW Calling Freq
3.885 AM Calling Frequency
3.799 DXpeditions SSB are frequently here
3.985 QRP SSB Calling frequency

40 METERS
7.000 - 7.010 CW DX Window
7.040 RTTY DX
7.050 XTAL Controlled Rigs
7.290 AM
7.065 DXpedition SSB USA split to 7.150 and above
7.005 DXpeditions CW are frequently here
7.110 QRP Novice/Tech CW Calling Frequency
7.171 SSTV
7.285 QRP Calling frequency
7.290 AM Calling frequency

30 METERS
10.106 QRP CW Calling frequency
10.110 -- DXpeditions CW are frequently here

20 METERS
14.025 DXpedition CW Freq -- Usually Split
14.060 QRP Calling frequency
14.080 DXpedition RTTY Freq
14.100 NCDXF Beacons
14.195 Rare DX & DXpeditions Frequently Operate SSB Here -- Generally Listening Up-Split
14.230 SSTV
14.285 QRP Calling frequency
14.286 AM Calling Frequency
14.336 MHz County Hunters when ever 20 is open and mobiles are around

17 METERS
18.075 DXpeditions CW are frequently here -- Usually Split
18.110 NCDXF Beacons
18.145 DXpeditions SSB are frequently here -- Usually Split

15 METERS
21.025 Rare DX & DXpeditions Frequently Operate CW Here - Generally Listening Up-Split
21.060 QRP CW calling frequency
21.080 RTTY DXpeditions are frequently here
21.150 NCDXF/IARU beacons
21.295 Rare DX & DXpeditions Frequently Operate SSB Here -- Generally Listening Up-Split
21.340 SSTV
21.385 QRP SSB calling frequency

12 METERS
24.895 Rare DX & DXpeditions Frequently Operate CW Here -- Generally Listening Up-Split
24.930 NCDXF Beacons
24.945 Rare DX & DXpeditions Frequently Operate SSB Here -- Generally Listening Up-Split

10 METERS

28.025 CW Rare DX & DXpeditions Frequently Operate Here – Split

28.060 QRP CW Calling frequency

28070.15 PSK-31 (offset -115 for USB)

28.080 RTTY Rare DX & DXpeditions Frequently Operate Here -- Split

28.1010 10/10 Intl CW Calling Frequency

28120.150 -- PSK31

28.120-28.300 Beacons

28.200 NCDXF/IARU beacons

28.380 10/10 SSB Intl Calling Frequency

28.385 QRP SSB Calling frequency

28.425 10/10 SSB Intl Calling Frequency – Another is 28.400

28.495 SSB Rare DX & DXpeditions Frequently Operate Here – Split

28.600 Old General Calling Frequency - Still used by Old Timers

28.675~28.685 SSTV Operating Frequency -- IARU Region 1

28.680 SSTV Operations USA/Canada

28.825 10-10 Backskatter Net - Paper Chasers Net

28.885 6M DX Liaison Frequency -- Listen here for 6 Meter DX opening announcements.

28.945 FAX Operating Frequency

29.000-29.200 AM Operations

29.300-29.510 Satellite Downlinks

29.520-29.580 Repeater Inputs

29.600 FM Simplex - Calling Frequency

29.620-29.680 Repeater Outputs

Here is our latest stealth tribander design!

CHAPTER VI   SIX METER DXING

6-1   CHAPTER CONTENTS

This section covers 6M DXing which can be considerably different than HF DXing due to the nature of the band.

6-2   EXPLODING THE BIG GUN MYTH

Over and over again the budding DXer will exclaim, "I'd like to work DX, but there is no way I can compete with the Kilowatts and beams". Certainly the big guns have an advantage but be assured you can work 6M DX with a modest station. One can work the six meter band with modest antennas at a reasonable height and low power if the band is open. Hopefully this book will give you a head start in that direction. Here are some examples. All are 6M DXers that I know:

Dennis has a 4 element beam and 100 watts and has worked 50 states in six months. Also currently has worked 13 DXCC countries and 313 grid squares.

Preston has an FT-100D and ATAS 100 mobile screw-driver antenna on a car port about 10 feet off the ground and worked 7 states in just 2 months.

Rod has a Cushcraft AR-6 Vertical on a 10 foot mast and has worked 49 states, 223 grid squares, and five DXCC countries in 3 years.

Don has been mobile on 6M for 6 months with IC-706 and a hamstick – has worked 25 states.

With a knowledge of propagation, operating techniques, some patience, and the proper time of the year, one can work VUCC (100 grid squares) quite easily. Some have done it in a contest weekend. Also it is not true that it is all "QSL ur 59(9) 73". Extended conversations can be had with many 6M stations both local and via skip.

6-3   FINDING DX

If you are new to VHF DXing, the USA Band Plans, Modes Of Operation Allowed, and Power Limitations – see Paragraph 6-14. The Calling Frequencies including DX Windows and common DX Frequencies are listed there also.

Unlike the HF bands, except for local contacts, the six meter band can be closed for long periods of time, so listening and tuning on 6M can be an exercise in futility. In the arena of HF DX, the old adage of listen, listen, listen was the way to find HF contacts but since 6M can be dead for long periods of time, a different strategy is order.

Some strategies to try are:

1. Tune to 50.125 MHz (the USA Calling Frequency) and leave the squelch open. The noise may drive you to distraction but this will avoid missing the weak ones. In the meantime, watch TV, read a book, search the web, or otherwise busy yourself as you may hear nothing or just a few locals all day (week) long. Needless to say listening in the off times of sporadic E or F layer skip will be like watching paint dry. See six meter propagation – Chapter 4.

2. Tune to 50.125 MHz and lightly squelch the receiver. This will keep the VHF noise from driving you goofy. This method is controversial as some say you may miss that very weak signal that is not strong enough to break the squelch. However in heavily populated areas, there always seems to be the 6M gurus who will spot any signal and call, thus breaking your squelch.

3. Don't hog the 50.125 MHz calling frequency by constantly calling CQ. At times one can hear several stations all CQing or making contacts in the pandemonium. Here is a tip that works well. Go to 50.125 and make a brief announcement "CQ CQ WA11XYZ EM45 going to 50.190. This will clear the calling frequency and possibly allow you to contact the weaker stations -- which is really what your looking for -- DX or new states or grid squares.

4. Don't rag chew with locals on 50.125 if you think the band is dead. It may open while you are keying down for your lengthy blurbs. A weak rare state or grid square may be underneath you.

5. Scan the SSB range say from 50.110 to 50.200 or lower if you are a CW operator. Hopefully your receiver will be free of birdies and noise spikes and allow an uninterrupted scan.

6. If birdies or other racket exists, try programming the memories every 5 kHz or so avoiding the birdies and noise sources and scan the memories.

7. Although 50.0-50.1 is designated as the CW only portion of the band, many CW ops will be in the 50.085 to 50.125 MHz range and sometimes higher.

8. In practice, CW activity cannot go much below 50.085 MHz or so, because international beacons fill most of the 100 KHz CW sub band. Even if activity starts around 50,090 MHz, CW activity is most likely to migrate up the band toward 50.125 MHz, and even higher in search of clear frequencies. There is simply nowhere else for CW stations to go. 50.090 MHz is the recognized calling frequency for CW.

9. Memory scan the six meter calling frequencies. CW at 50.090, SSB at 50.125, FM at 52.525, and AM at 50.400 MHz. For good measure, include the DX calling frequency at 50.110 MHz.

10. Don't overlook the FM portion of the band for DXing, See FM calling frequencies at the end of the Chapter. Growing widespread use of 50.300 FM has recently produced many interesting and unexpected DX contacts throughout North America. It is ideal for those operators whose antennas favor the low 'DX' end of the band. Be sure to include it into memory and scan groups wherever applicable.

11. The same is true of the AM mode which is gaining popularity, with the calling frequency at 50.400 MHz. If there's no local AM work in your area you might check for it during a good band opening.

12. Check the DX packet clusters and the DX Telnet, numerous 6M spots are made there. But unlike HF, you may not be able to hear the station reported due to the nature of the band. It is not uncommon for a station a few hundred miles away to make a DX spot but you may have no propagation and can't hear the station. But the cluster can be a good source of band openings. Spots from the other side of the country may have little meaning, unless the spot reports propagation in the general vicinity of your grid square. DX Packet Cluster and Telnet links are at URL: http://ac6v.com/dxcluster.htm

13. Check the Beacon range 50.06-50.09, often one can hear a beacon but not hear any stations from that locale, try a directive CQ and sometimes this will get a response. Six meter operators can be scarce during the work week and a good deal of time from the less populous states such as Wyoming and Rhode Island. Weekends and contests usually have more activity from these states. For International Beacons from 50.00 to 50.09 MHz see URL: :http://ac6v.com/beacons.htm

14. 28.885 MHz Liaison. During the previous solar cycle, 6-meter operators around the world used 28.885 MHz as an information and liaison frequency. There was nothing like getting reports on conditions first hand or making immediate arrangements with specific stations to listen on 50 MHz.

15. Even with the Internet, it is likely that the 6-meter liaison frequency will still be useful, especially for stations not yet wired into the Web and e-mail.

16. On 28.885 MHz, call "CQ 6-meter activity Europe" (or some other specific geographic region) if you like, or listen for DX stations seeking North Americans. Once you find someone on 28.885 MHz, move up or down in 5 KHz steps to leave the liaison frequency open for others. The 28.885 MHz liaison frequency is also a great place to monitor, as a good deal of gossip, news and timely band reports get passed here.

17. If your local 2M repeater has several six meter hounds, make a brief announcement of a six meter band opening. Or if this bothers the 2M crowd, set up a 2M simplex frequency for your local group.

18. If you have lots of time and patience, periodically tune the band for activity. Sporadic E can occur at just about any time, but the hour-to-hour variation in Es propagation during the summer months is typically with peaks occurring between 10 A.M. and noon, local time, and again from 6 P.M. to 8 P.M. Es propagation is generally a daytime phenomenon during the summer months, decreasing rapidly after local sundown.

6-4   WORKING THE SIX METER BAND – NON CONTEST

Exchange essential information quickly. Unlike the HF bands, contacts can be very short, lasting just a few seconds due to shifting propagation. This can be particularly true for multi hop contacts. Thus it is imperative to quickly nail down the callsign, grid square (4-digit only), and a signal report, in that order. On the next go around, the name, city and state are typically exchanged. Don't start to rag chew or give other information until these essentials are out of the way. Many a broken contact has occurred because the stations didn't get the required info across and 10 seconds later the band went kaput. If you and the contacted station are amenable to a rag chew, by all means continue unless you suspect a short band opening and want to hunt for other states and grid squares.

Long conversations can be held at times depending on propagation. A 15 minute QSO is not uncommon on single hop sporadic E contacts. But sporadic E propagation can be very fickle – very strong and seconds later – poof gone. F layer contacts can be much like 10 meters, the band may stay open for long periods.

Listen for US District Callsigns. You should have a US district map handy so you will know what states and areas you are hearing. See appendix A-9 for a USA District map. But a W6 can live in W4 land so districts can be misleading. Often you will hear a station in QSO give their callsign and grid square and you may want to know their state and not listen to an extended QSO. In this case you need an aid to show what state the grid square is located in. Likewise, you may only receive or hear a grid square. See URL: http://ac6v.com/opaids.htm#GRID for maps and grid square aids. But with some maps, it is difficult to determine the state borders. Programs to show grid squares are covered in Chapter 3 and the Appendices, A6, A19, A20

Listen for Grid Squares. The Maidenhead grid-square system, formalized at a VHF meeting in Britain in 1980 and adopted world-wide by the International Amateur Radio Union in 1985, is almost universally used as a locator system by VHF, UHF and microwave operators.

The Maidenhead system divides the world into 32,400 squares, each 2 degrees of longitude by 1 degree of latitude.

There are larger "fields" of 100 locator squares each, and each square is divided into smaller "subsquares." For most purposes, knowing your 2 degree by 1 degree square is sufficient.

The Maidenhead Grid Square locator system has been long used by the VHFers, but more and more HFers are asking and chasing these. For your grid square, locators and maps (USA and World-Wide), see URL: http://ac6v.com/opaids.htm#GRID

Another compelling reason for using grid squares is because many awards (ARRL and those from six meter organizations) are based on the number of grid squares worked.

A neat program for this is DX Atlas URL: http://www.dxatlas.com/ By VE3NEA. With a map displayed of the area desired, do a CNTRL G, then type in the grid square, a red 4-point locator will appear on the map so you can determine the state and location within the state. With this program, the state outlines are very clear. Where the grid square is close to state borders, using the zoom function will enlarge the grid square to help determine a more precise location.

In the case where the grid encompasses 2 or more states, you can use the RAC, Buckmaster or QRZ callbooks to determine the caller's state and city. (I have it at the ready in my computer along with DX Atlas and a Logging program).

Other useful tools to coordinate QSO's are the 50 MHz prop logger (http://dxworld.com/50prop.html) as well as the chat area at www.dxer.info. Sometimes you only hear the callsign and want to know the city, state and grid square. Use the callbooks for this and hope the guy/gal didn't move.

Another valuable source for grid squares is the Repeater Map Book By Artsci. It shows USA grid squares for each state in map form. Also they have the major highways, cities, and repeaters on the map. URL: http://www.artscipub.com/mapbook/

DXCC entities can be far and few in between, be sure to hang in the pileups to get them, you may not hear another like it for a long long time.

Log All Contacts. And sooner or later you will want to have in the log, callsign, state, and the grid square as well as mode, time (UTC) date and RST. You will receive QSL cards, particularly if you live in a rare grid square. Then if you decide to apply for VUCC or Worked All States, you will have a record of all your contacts. Required logging information and QSL card formats are given in Chapter 9.

Tune the entire band, you may find a station well up from 50.250 MHz. Sometimes two locals will QSY well up in the band for a somewhat private QSO, but you can call them when the QSO is finished. Break maybe if you are desperate for the contact – your choice.

<u>**Switch To CW**</u>. Sometimes it becomes almost impossible to get the information on phone, so you can switch to CW. Don't be too concerned with your code speed, in general the 6M CW ops are quite a bit slower than 20 M types. If the code is too fast – send a QRS and most ops will accommodate you.

<u>**Double hop Sporadic E contacts**</u> are very desirable as these openings are not as common as single hop. When double or multihop openings occur, it is an opportunity to work US states out to 2500 miles distant or DX. Listen for the weaker stations with lots of fading.

<u>**ESP contacts.**</u> Maybe Expected Sensory Perception. The faint of heart will exclaim, "I can barely hear the DX station". Well that is good enough. So what if it takes six overs to get the reports across – jump in. On several memorable occasions, the Little Gun with good ears has been the last, or one of the last, to work the DX before they shut down or went to another band.

The human brain seems to be capable of a certain amount of weak signal processing. When listening to a very weak signal for a long period of time, the brain seems to be able to filter out the background noise, and the weak signal becomes discernable.

<u>**Does power and gain antennas matter?**</u> Well with good F skip or single hop sporadic E, a few watts and a dipole or vertical will work surprisingly well. With poor propagation and multi hop, going QRO can get you above the noise level. And of course a directional high gain antenna will let you hear the weak ones. Hence "The Magic Band"

<u>**Antenna polarization**</u> is not a major consideration for DXing as there is a great deal of bending of the RF wave as it bounces around the ionosphere and polarization can be anything – sometimes vertical, other times horizontal, or something in between. For line of sight contacts, polarization is important as cross polarization attenuation can be as high as 20dB.

The choice is difficult as the big guns are likely to be horizontally polarized and the mobiles use verticals. By long standing tradition, some areas may favor one or the other polarizations, check with the local gurus. Some stations have both horizontal and vertically polarized antennas.

6-5   WORKING SPLIT

When the band opens to other countries, the competition can be fierce, so some DX stations may operate split. The figure below shows a typical split operation

In the example above, the DX station transmits at 50.110 MHz, but listens for calls up band from 50.115 to 50.120 MHz. Use your VFO A to listen to the DX station at 50.110, set your VFO B to transmit somewhere in the 50.115 to 50.120 MHz range. On receive you be tuned to the DX frequency and on transmit you be on the VFO B frequency where the DX station is listening.

6-6 WORKING CONTESTS

Contests bring lots of stations out of the woodwork and if there is a band opening, many quick contacts can be made. You do not have "to be in the contest" to participate-- that is submit a log to the sponsoring organization. For a good contest calendar and rules – see URL: http://ac6v.com/contestlinks.htm

6-7 QRP OPERATION

Because signals can be so strong at times on the six meter band, low power operation is popular. Running 5 Watts or less will net a lot of contacts if the band is open. You can build a 6M QRP rig or just use your power level control reduced to 5 Watts, Working all states QRP is achievable on the six meter band, but requires lots of patience and determination as all QRP operations do.

6-8 CW OPERATION

If you are new to CW DXing, you may want to read Appendix A1, CW operating procedures before continuing. For CW, the beginner can get shaky and even send their call pretty raggedly. Get a memory keyer until you get used to it all. Some of the more exotic keyers can be used to memorize your call, report, increment serial numbers, etc. Just push buttons!

For CW contests – jump in – with computer practice, you should be able to recognize your call at any speed – so you'll know when they come back to you. Just reply TU ur 5NN and the report (grid, etc.). Same with non-contest pileup DX just 5NN and grid square.

6-9 DX PACKET CLUSTERS, TELNET AND THE DX SUMMIT

Before repeaters, DX brethren would have a network on VHF simplex to inform each other of DX. This was supplemented with the infamous "one-ringer" – a phone call at 2:30 in the morning alerting the conked out DXer to get on the air. Buddies kept lists of each other's needs and the "one-ringer" system aggravated many an XYL. In the 70's exclusive DX 2M repeaters came into being, all QSO's were more or less DX oriented and DX spots took priority.

In the 80's, a new phenomenon came in to being – the DX Packet Cluster. Using computers and VHF packet radio with AX.25, one station is set up with software and is linked to one or more other stations who have installed the same system. These nodes when connected are called a cluster. DX announcements propagate thru the cluster to all users. Later the clusters interconnected to the next outlying one and soon statewide DX Packet Clusters were in operation. For 6M DX spots and openings – use the local VHF packet cluster or you can also use telnet on the internet --- URL: http://www.ng3k.com/Misc/cluster.html

Samples of DX Packet Cluster announcements today are:

Some Packet Cluster abbreviations:
CBA – Call Book Address
QSX xxxx - The split frequency where the DX is listening
EU Only – DX is listening for Europe only
NA (SA) Only - DX is listening for North America (South America) Only
NIL – No Signal Heard
UP – Split Operation - DX is listening UP (usually 5 to 10 kHz SSB, 2 to 5 kHz CW)
MNG (MGR) – QSL Manager of the DX Station
VIA - QSL Manager of the DX Station
DIRECT – DX is listening and transmitting on the same frequency
SIMPLEX - DX is listening and transmitting on the same frequency
CLG (CQ) – DX is calling CQ
HR = Here
WKD (WRKD) – Worked (A Brag Spot)
(NA) (SA) (AS) (OC) (EU) (AF) (AN) continents - North America, South America, Asia, Oceania, Europe, Africa, Antarctica
AF-032 - IOTA Number for continents e.g., AF-032 is the Islands On The Air Number for Zanzibar Island off the African Coast

A "busted" spot is one in which the callsign sent to the cluster is wrong - not the callsign the DX station is actually using. Was a typo or the spotter miscopied it.
WARC – The WARC Bands are 30 meters, 17 meters, and 12 meters
/B – Beacon

6-10 WORKED ALL STATES

Work and confirm all 50 states in the USA in any combination of bands/modes will earn the basic certificate. Specialty certificates are issued for a variety of different bands and modes such as Satellite, 160 meters, SSTV, RTTY, each VHF band. Available endorsements include SSB, CW, Novice, QRP, Packet, EME, and any single band except 30 meters.

6-11 WORKED 100 GRID SQUARES (VUCC)

The ARRL offers an award for working and confirming contacts with 100 different grid squares. When seeking this award you will want to have a list or spread sheet made up of the grids you have worked and confirmed.

VHfers don't seem to send QSL cards as much as HFers do – so if you have worked a particular grid but have not received a card in a reasonable period of time – then work them again as a matter of an insurance contact.

All about QSLing, domestic and DX is covered in Chapter 9.

6-12   DXCC

The ARRL offers an award for working and confirming 100 DXCC entities. Rules can be found at URL: http://www.arrl.org/awards/dxcc/

6-13   OTHER AWARDS

Several other organizations offer awards, use the search engine Google and search for "six meter awards"

6-14   SIX METER FREQUENCY ALLOTTMENTS

For the USA, Amateur usage of the six meter band is officially set by the FCC and is simply:

All Amateurs except Novices:
50.0-50.1 MHz: CW Only
50.1-54.0 MHz: CW, Phone, Image, MCW, RTTY/Data
Maximum 1500 Watts PEP

To elaborate on the modes, phone can be AM, FM or SSB and all three are used on six meters. Other phone modes could be used but are seldom heard such as double side band suppressed carrier. The single side band convention above 30 meters is upper sideband (USB) although one can operate LSB and be within the rules, but since everyone operates USB, there is little reason to do this.

Note that CW can be used over the whole band, and on occasion, you may hear CW on the band portions typically considered as the phone portions.

A typical use would be two stations are in SSB contact and the band fades where SSB just isn't feasible any longer and the stations will switch to CW for the QSO completion. Or two stations are in phone QSO, and one wants a CW contact, perhaps because they are working all states or DXCC in the CW mode, so they quickly switch to CW for the contact. MCW - Telegraphy by Morse code with on-off keying of a tone modulated carrier, seldom used on the ham bands these days.

Data modes are growing in popularity including RTTY, PSK31, AMTOR, PACTOR, G-TOR, CLOVER, HF PACKET, HELLSCHREIBER, MT63, THROB, MFSK16, and others – See URL: http://home.teleport.com/~nb6z/frame.htm

6-15   USA RECOMMENDED SIX METER BAND PLAN

To bring order out of chaos, the ARRL and several other six meter groups have recommended a band plan which almost all six meter operators now follow.

50.0-50.1	CW, beacons
50.060-50.080	beacon subband
50.1-50.3	SSB, CW
50.10-50.125	DX window
50.125	SSB calling
50.3-50.6	All modes
50.6-50.8	Nonvoice communications
50.62	Digital (packet) calling
50.8-51.0	Radio remote control (20-kHz channels)
51.0-51.1	Pacific DX window
51.12-51.48	Repeater inputs (19 channels)
51.12-51.18	Digital repeater inputs
51.62-51.98	Repeater outputs (19 channels)
51.62-51.68	Digital repeater outputs
52.0-52.48	Repeater inputs (except as noted; 23 channels)

52.02, 52.04	FM simplex
52.2	TEST PAIR (input)
52.5-52.98	Repeater output (except as noted; 23 channels)
52.525	Primary FM simplex
52.54	Secondary FM simplex
52.7	TEST PAIR (output)
53.0-53.48	Repeater inputs (except as noted; 19 channels)
53.0	Remote base FM simplex
53.02	Simplex
53.1, 53.2, 53.3, 53.4	Radio remote control
53.5-53.98	Repeater outputs (except as noted; 19 channels)
53.5, 53.6, 53.7, 53.8	Radio remote control
53.52, 53.9	Simplex

CW – Although 50.0-50.1 is designated CW only, most often this is where beacons will be heard. These are like light houses of the air ways and are operated by Amateurs world wide. They are a great way to check for propagation. USA and World Beacons can be found at URL: ac6v.com/beacons.htm

CW Non-Beacon is most commonly heard in the 50.080 to 50.100 MHz range

SSB operation is generally in the 50.100 to 50.300 MHz range although during the big band openings many SSB signals can be heard above 50.300.

AM operation is usually from 50.300 to 50.600 MHz although this range is for any mode of operation. Data modes are found in the 50.600 to 50.800 MHz. range

From 50.8 to 51.0 MHz with 20-kHz channels is the portion allotted for radio control of non amateur hobbies such as boats, aircraft, and even rockets. Another segment for R/C is 53.5, 53.6, 53.7, 53.8 MHz. Best stay off these so as not to trash someone's $500 model airplane!

A rather odd allotment exists at 51.000 to 51.100 MHz. Dubbed the Pacific DX window; it was apparently included for Pacific area stations for regular use. But many Pacific DX operations are typically found at the SSB DX window 50.100 to 50.125 MHz.

The FM range for six meters is from 51.100 to 54.000 MHz... Included are voice and digital repeater inputs, repeater outputs, and simplex frequencies. See Paragraph 2-6. The best bet for finding repeaters is the ARRL repeater guide and the on-line repeater guides at URL: http://ac6v.com/repeaters.htm

6-16 CALLING FREQUENCIES

Calling frequencies are generally accepted ham agreements or ARRL recommendations as to where to make a general call for the various modes and activities. If after tuning the range for the mode you are using, no activity is heard, the calling frequencies are a good place to call CQ. After a contact is made, it is best to move off of the calling frequency to continue the QSO so that others can use it for its intended purpose. Too many folks are camping out on the calling frequencies, calling endless CQ's or rag chewing, making it impossible for the listeners to catch some DX.

50.058 Fists CW Club Calling Frequency
50.090 CW Calling Frequency
50.110 DX Calling Frequency (USB) Usually Non-USA Stations Call Here.
50.115 DXpeditions Frequency operate CW and SSB here
50.125 National SSB Simplex Frequency (USB) Lots Of USA Hams Call Here For Local and DX
50.260 is the WSJT Meteor Scatter calling frequency in the USA -- 50.270 is FSK Meteor scatter
50.300 FM Simplex Calling Frequency (West Coast)
50.385 PSK31 Calling Frequency
50.4 National AM Simplex Frequency
50.7 RTTY Calling Frequency
51.910 FM Internet Linking
52.525 National FM Simplex Calling Frequency

TUNING IN A NUTSHELL

Beacons - tune around 50.060-50.080

CW - tune around 50.090 and call CQ at this frequency

SSB USA - tune around 50.125 (USB) and call CQ at this frequency

SSB DX - tune around 50.110 (USB) and call CQ DX at this frequency

AM - tune around 50.4 and call CQ at this frequency

PSK - tune around 50.385 (USB) and call CQ at this frequency

RTTY - tune around 50.7 (LSB) and call CQ at this frequency

FM - tune around 52.525 and call CQ at this frequency

But don't overlook the lower CW frequencies and the DX windows

You might want to put these in the transceiver's memory for quick access as well as the data for local repeaters (offset, PL, etc). Also see Chapter 5, Working DX, for more tuning tips and finding DX.

6-17 TYPICAL SIX METER FM BAND PLAN

Following is a typical USA six meter band plan for FM, but may vary from area to area, check with your local coordinating group.

DUPLEX VOICE FM REPEATERS (Standard Pairs)
39 Repeater Pairs, 20 KHz spacing, 1 MHz - in/out

52.010 - 53.010	52.030 - 53.030	52.050 - 53.050	52.070 - 53.070
52.090 - 53.090	52.110 - 53.110	52.130 - 53.130	52.150 - 53.150
52.170 - 53.170	52.190 - 53.190	52.210 - 53.210	52.230 - 53.230
52.250 - 53.250	52.270 - 53.270	52.290 - 53.290	52.310 - 53.310
52.330 - 53.330	52.350 - 53.350	52.370 - 53.370	52.390 - 53.390
52.410 - 53.410	52.430 - 53.430	52.450 - 53.450	52.470 - 53.470
52.710 - 53.710	52.730 - 53.730	52.750 - 53.750	52.770 - 53.770
52.790 - 53.790	52.810 - 53.810	52.830 - 53.830	52.850 - 53.850
52.870 - 53.870	52.890 - 53.890	52.910 - 53.910	52.930 - 53.930
52.950 - 53.950	52.970 - 53.970	52.990 - 53.990	

DUPLEX VOICE FM REPEATERS (Standard Pairs)
15 Repeater Pairs, 20 KHz spacing, 500 KHz - in/out

51.200 - 51.700	51.220 - 51.720	51.240 - 51.740	51.260 - 51.760
51.280 - 51.780	51.300 - 51.800	51.320 - 51.820	51.340 - 51.840
51.360 - 51.860	51.380 - 51.880	51.400 - 51.900	51.420 - 51.920
51.440 - 51.940	51.460 - 51.960	51.480 - 51.980	

DUPLEX FM PACKET/DIGITAL REPEATERS
5 Repeater Pairs, 20 KHz spacing, 1 MHz - in/out

52.610 - 53.610	52.630 - 53.630	52.650 - 53.650	52.670 - 53.670
52.690 - 53.690			

4 Repeater Pairs, 20 KHz spacing, 500 KHz - in/out

51.120 - 51.620	51.140 - 51.640	51.160 - 51.660	51.180 - 51.680

VOICE FM SIMPLEX FREQUENCIES
12 Simplex Channels

51.500	51.520	51.540	51.560	51.580	51.600
52.490	52.510	52.525*	52.550	52.570	52.590

* 52.525 is the National FM Voice Simplex Calling Frequency

PACKET/DIGITAL FM SIMPLEX FREQUENCIES

53.490	53.510	53.530	53.550	53.570	53.590

6-18   OPERATING AIDS

Log Book – Log all contacts in UTC time, be sure to include grid square and state or country. If you received the grid square in the exchange but not the state - look them up on QRZ.COM. Their listing also includes the grid square under detailed information. The Radio Amateur Book has this information as well. However some folks move and neglect to inform the FCC or are a mobile (rover) station, so ask for all information if time and band conditions permit.

Grid square Map. The grid square map from ICOM is excellent. URL:
http://www.icomamerica.com/ Click on AMATEUR, then U.S. Grid Square Map. . Print it out in landscape orientation. X them out as you work them. This provides a quick reference to the grids you have worked.

DX Atlas. URL: http://www.dxatlas.com/ Has menus that allows you to type in a grid square number and the program places a cursor on the map presentation allowing you to determine the US state or country/province. Also gives distance, bearing, and prefixes.

Radio Amateur Book. Often you will hear a call sign but not sure of the grid square or state. Having the RAC handy will let you obtain the information quickly. Also you can use QRZ.COM.

DX Packet and Telnet Clusters See paragraph 6-9.

Scratch Pads – The six meter band can be very volatile so you may hear a station callsign (or grid square) but can't work them. So rather than clutter up the logbook, use a scratch pad notes until everything is confirmed.

Beacons – For International Beacons from 50.00 to 50.09 MHz see URL:
:http://ac6v.com/beacons.htm

Computer. Great for logging which allows for easy electronic mailing for awards – see ARRL: Log Book of the World URL: http://www.arrl.org/lotw/ Also for maps, telnet, etc

---

CHAPTER VII   DX SECRETS

7-1. THE DX SECRETS OF THE AGES

UPON PAIN OF HIGH VSWR, ALL WHO ENTER HERE MUST AVOW NEVER TO REVEAL THE DX SECRETS OF THE AGES TO A NON-BELIEVER.

NO! I'm In Poland!

NOTE: Throughout this section, reference is made to the Callor – that is the one running the pileup. The callee is the poor guy trying to get a contact.

7-2. FINDING DX

1. Listen, Listen, Listen

2. Be aware of propagation, check the SFI, A, and K indices.

3. Use the NCDXF/IARU Beacon System to determine propagation paths. You can read a dozen pages on propagation paths during the seasons, sunspot highs and lows, etc., but you'll get real time propagation data with the NCDXF Beacons.

4. Try Miniprop by W6EL for propagation predictions, but don't be surprised if you can work an S79 when Miniprop says no way. Try WinCap or IONCAP programs also.

5. Use your local DX Packet Cluster, telnet, or OH2AQ for DX Spots. These are so efficient that worldwide pileups can develop in seconds, best that you be on the "spot circuit". Better still, tune, tune, tune, find em first, work em, spot em – In that order, never spot until you've worked. Don't get addicted to the cluster, however -- there is a lot of DX out there that will only respond to CQs because they hate pileups, want to ragchew, or don't have the time for working too many stations. Also valuable is DXMon http://www.benlo.com/dxmon.html excellent and free.

6. From the DX spots and your own experience, determine when the bands are open for DX. For 10 meters there might be a nice opening to Europe in the morning, but fades away later in your day as darkness falls on Europe. Switching to 15 meters might well find a path into Europe, and later 20 meters.

7. Be aware of the frequencies that DXpeditions and rare DX operations tend to operate – see Chapter 5.

8. Do get your code speed up. Some DXpeditions or operators operate CW only and you don't want to miss out on an extremely rare island that is visited once every 15 years. CW DXing generally finds a smaller pileup than phone, as many DXers do not work CW. The exchange is short and with a little practice you can work CW expeditions.

9. Be prepared to operate ALL modes ALL bands CW, SSB, RTTY, 6 meters etc. – DXpeditions do operate RTTY and some 6 meters and that thins out the crowd!

10. The 12 meter and 17 meter bands are not usually as busy as 10, 15, and 20 meters. Maybe because many folks only have tri-banders and the popularity of 10 meters for techs. Also no contesting is allowed on these bands. Get on 12 meters and 17 meters, many a time you can hear Africa or the Marianna's calling CQ with no takers.

11. If you have only a small vertical and restrictions, remember that 40 and 80 meters are night bands when it comes to DX and a 12 foot 40 meter vertical is very very inefficient. Also the trapped verticals are inefficient and very narrow banded. So try making a 32-foot vertical out of a fishing pole and a wooden dowel, using zip lamp cord as the radiator. Cut two or four more 32-foot sections and use as radials, preferably elevated on a garden wall. This will make a nice light-weight antenna that can be erected at night and wow; you will be surprised with the performance compared to the shorty verticals.

12. For 40 meters - better still if you can build a 40-meter vertical phased array with lots of radials. But do get an efficient antenna on 40m. I couldn't work Macquarie Island on the upper bands, but my fishing pole antenna made the trip on 40 meter CW (600 Watts), very early in the morning. For 80 meters, height is a problem, so try base loading (or better still, center loading) the 40-meter vertical. Make the loading coil out of heavy-gauge wire with a big diameter coil form. Check out base loading coil design in the ARRL publications.

13. Read all the latest DX Bulletins on the Internet, great way to learn what DX is on the air along with operating frequencies in many cases. NG3K's pages are the place to start.

14. Check for the operating habits of the desired DX. Folks tend to come on the air at more or less the same time and frequency – use the DX Summit for spot information. Plan to be all tuned up and ready when the EK comes on every Monday night at 0330Z. In fact give him a call then, but that will attract attention and the big guns may beat you out.

15. If a DXpedition has published or announced the operating frequencies for the DXpedition, have ALL of them in the radio memory. When the DX announces, I'm QSY, going to 12 meter CW – you will be ready. Have the split frequencies all set up also.

16. It is essential that you can recognize prefixes or have a Prefix Cheat Sheet handy – See Appendix A10. When an HM#DX comes on, you will know where to point the beam and realize you have just found the rarest of the rare. HM is North Korea!

17. Antenna polarization is not a major consideration for DX as there is a great deal of bending of the RF wave as it bounces around the world. Vertical antennas are prone to noise however.

18. Be aware of time differences around the world, e.g., Europe is 5 to 8 hours ahead of the USA.

19. Gray Line Propagation -- use the twilight line (also known as gray line), DX can happen best along the line between day and night. Look for a time when the area you want or your area is in this twilight time of day. Some very nice programs for this – see Chapter 3.

20. On phone, listen for accents; you will be able to distinguish a W6 from a VK easily.

21. Listen for the distinctive DX sound, watery, polar flutter, lots of QSB, weak signal, and that far-off station "sound", a VU doesn't sound like a W6, that's a given.

22. On CW listen for drifting signals and chirp, often indicative of older equipment or homebrew rigs - both common in some countries. Many countries have poor power line regulation, resulting in chirp and drift.

23. Tune the entire band, you may find a TF (Iceland) at 28.700 during a contest and a ZD8 at 21.400! For CW, check off the beaten path e.g., 28.070 and above. Don't forget the novice bands, some DX stations will operate there and work whoever calls.

24. Tune the bands after they "go out" because they sometimes come back up. One night after the "band went out", a fairly rare Asian station A4 called CQ on 17 meters late at night -- no one answered – 'cept me.

25. You hear a massive pileup at 28.505 MHz but you don't hear a DX station in there or can't figure out what is going on. You may not have propagation where the east coast does OR the DX is working split and not spreading the callers out. In the first case, don't ask on frequency, check the DX Summit on the Internet – it will probably reveal who the DX is. In the second case, an example, DX11DX is on 28.495 MHz and telling callers "listening 28.505 MHz". Since most split operations are listening up, tune below the callers to try and find the DX. Usually a cop will be there (unfortunately) alerting simplex callers "He's listening up".

26. You hear a massive pileup spread out from 28.505 MHz to 28.510 MHZ. This is a tip-off to a split operation. An example, DX11DX is on 28.495 MHz and telling callees "listening 28.505 to 28.510 MHz". Since most split operations are listening up, tune below the callees to try and find the DX. Usually a cop will be there (unfortunately) alerting simplex callers "He's listening up".

27. Use the DX Nets if you have to or want to, but be aware "True Blue DXers" look askance at the practice. And don't mention it at your first DX Club meeting. Even some "true blue DXers" will come on a DX Net when the P5 shows up there (or only there).

28. If a possible new country comes on the air, e.g., Sealand. Work em now - no matter how dumb it sounds that they will become a new one. I missed Scarborough Reef that way!

29. Be aware of what is rare, medium rare, and well-done DX. See the top most wanted list on the Internet and Appendix A12. Early on, I missed a TT8 this way, when I didn't know how rare it was!!

30. A DX station is questioned as being a slim, WORK EM, WORRY LATER. Beware of DX spots on the cluster announcing slim, pirate, bootlegger, etc. Many a time this was just a supposition, and it was the real McCoy!

31. For Six Meter Propagation – see Chapter 4. For Six Meter Dxing – see Chapter 6. On the DX packet clusters or telmet, someone is always listening. Great way to spot openings and of course make spots yourself. Tune the rig to 50.110 MHz and lightly squelch, good way to have a 6M alert.

32. Yeah "QRP is for sissies" – so some one sed, or "I didn't get to the top of the DX chain to work QRP". But don't overlook the QRP listening frequencies – lots of DX there. Try CW-- 3.710, CW--7.040, SSB --7.285, CW--7.035 (QRP-L), CW--7.110, CW--7.112 (NorCal crystals), CW--10.106, CW-- 10.116 (QRP-L), CW-- 14.060, SSB --14.285, CW--18.096, SSB --18.130, CW--21.060, SSB --21.385, CW-- 21.110, CW--24.906, SSB --24.950, CW-- 28.060, SSB -- 28.885, CW--28.110, SSB --28.385. Just turn down your RF Power Control.

33. Work DX at odd times, their holidays, Super Bowl time, 1:30 in the morning, lunch hour if you gotta work. Many a DXer gets sick as a dog when a ZD9 comes on, exercising their "flex time off". When 20 meters is open all night, the pileups diminish considerably by midnight.

34. If you speak other languages, call CQ DX in that language, boy does this work! Or in a pileup, call the DX in their native language if you can speak it moderately.

35. If you have language difficulties on phone or CW, use Q Signals as much as possible. There are Q-Signals for many situations that experienced CW operators will know.

36. Don't pass up the IOTA stations (Islands On The Air). They count for the parent countries, e.g., TA0/IT9YRE AS-154 counts as Turkey.

37. Get a 10/10 number – worked some rare ones with a sure QSL that way. See URL: http://www.ten-ten.org/

38. Looking for a way to E-Mail a DX Station for a contact? Try QRZ.com. For example, Bahamas can be found by going to QRZ.com, then click on find by name, and type in Bahamas. Yes it finds by keyword as well, and five or six Bahamas stations are listed. Don't know that any of these are on HF, but what have you got to lose?

39. Check out the World Wide clubs and organizations on the Internet, many have schedules for nets and E-Mail addresses. Example, Pitcairn island has on-line schedules.

40. Check out the Ham Chat Channels on the Internet. Hams from all over the world check in and you can ask about their HF habits.

41. Check the county hunters frequencies, some nice DX shows up there. Better still, go mobile to a rare county, and the DX will call you.

42. Make a schedule by E-Mail or snail mail.

43. Get a 2x1 or 1x2 call, seems easier for the DX to pick em out, maybe not.

44. Get a call with single syllables – double U A six double U double U double U WA6WWW ain't gonna command much attention.

45. Get a call with lots of syllabance --- K Six S S S will cut right through. Consonants frequencies are from about 2000 Hz upwards and carry a lot of the intelligence of a signal.

46. Get a rare prefix and work the WPX contest – DX will call you!

47. Join a DX club – when they go on a DXpedition, they work Club members first or will recognize your call.

48. Join a DX club – your new buddies will give you a one ringer at 2:30 in the morning when the P5 comes on (after they have worked em)

49. Join a DX club and commiserate with the Honor Rollees, they didn't work 350 countries by calling CQ DX.

50. Listen, Listen, Listen

7-3. CALLING DX (NON - PILEUP OPERATION)

51. Here again know when the DX will propagate to you. See above.

52. Many DX stations will be operating in the early evening after they get home from work. Or early in their morning. Remember Europe is five to eight hours ahead of the USA.

53. Adjust your operating hours to match the DX leisure hours. Some mighty fine DX can be worked at 3:00 AM. Work em – go back to bed (with a smile)!!

54. Old adage says, if you can hear the DX, you can work them, don't play "Little Gun", call em. The DX might well be operating with 50 Watts and a dipole and is very weak, but if you can hear the DX, chances are the DX will hear you.

55. Calling CQ DX can be an exercise in futility, but can be a surprise on occasion; especially when the bands are quiet and you have a hunch the band is open to "deep central Asia".

56. If you are a moderate gun or a big gun, try the directive call "CQ Deep Central Asia". I have heard some surprising replies as well as a guy in Modesto who quips, "You are also bombing into Modesto, CA, OM".

57. A myth is you can't rag-chew with a DX Station. Listen for them calling CQ, unless they are very rare, many will engage in an extended QSO, learn some of their history, ask questions, etc.

58. Sure-fire conversation getter. Hello CQ DX, CQ DX from W$xxx, no QSL's just QSO's. Have generated pileups this way. Some of the old timers hate the "send me a QSL" routine.

59. On phone, for non-English DX, speak slowly using standard phonetics or DXers phonetics i.e., America, Brazil, Canada, Denmark, England, etc. Don't use cute phrases like WA6 William Tell Overture. South and Central America prefer Brazil, Colombia, Chile, Guatemala, and Lima. See Appendix A4 for a complete list of phonetics.

60. If at one time you were on CB radio – that is fine, however leave your CB lingo back at the 11-meter band. Such phrases as "The personal is", "hitting me with six pounds", 10 codes, police phonetics, and other CB lingo grates against a lot of older and non-cb hams. When in Rome, do as the Romans. Likewise on VHF FM there is no reason to use Q-Signals (QTH, QSL, etc.), lots of non-HFers don't understand these. Why not just say I'm at home, instead of at the home QTH??

61. When a DX station announces they will QRT or QSY, throw in your call, especially if you have met or know the operator. P40V always comes back to me!

62. Rich KY6R reports a neat trick. . I heard YI1BGD reply to someone else's CQ, so I decided to call CQ 3 KC's up from that QSO, and he replied to my CQ. He must have been tired of the pileups and instead made a few QSO's by answering CQ's. I had been chasing him for weeks on 40 and 30 CW during the gray line in January, and he was always about 559 at best with heavy pileups. Anyway, this confirms that the most important ham "gear" is your ears (and mind).

63. Strange how many DX stations announce QRT but come back in a few minutes (potty break?) Hang in there.

64. The DX announces "Back in 10 minutes", maybe -- but better tune up and down the band, they may not want to come back to this zoo of a pileup. Happens often. Put his current frequency in radio memory and tune around or check the other bands.

65. Listen to pileups where DX stations are piled up on someone in your hemisphere, i.e., Europe piling up on a HP Panama station. Put the frequency in memory, when the HP secures, try calling the needed Europeans.

66. Announce your state if rare – AC7V in Wyoming will get em every time!

67. Europe is rolling in and your beam is pointed east, better swing it around and listen as an XZ Myanmar may be coming in off the back of the beam! Happened to me on 20 meters during a contest. But then again I have a vertical.

68. In a contest, you hear a weak CT3 from Madeira calling CQ contest with no takers, but with the 20 over S9 signals all around and QRM galore, you figure no use. Wrong – he is running a dipole with 50 Watts and comes right back.

69. Don't be timid with the ESP contacts. So what if it takes six exchanges to get the reports across – amazing how patient DXers are with the weak ones. The faint of heart will exclaim, "I can barely hear the DX station". Well that is good enough. Example, it is late late on 20 meters, the east coast is asleep, and the west coast is hearing watery weak polar flutter signals at best. R1MVI is CQing – maybe every 3rd syllable is understandable – no one responds. Needing this one in the worst way, I call back AMERICA CANADA SIX VICTORIA, it took a half dozen overs, but we finally made a two-way QSO under the most adverse of conditions.

70. In weak DX contacts on phone, use OVER, OVER when you turn it back to the DX. The QRM and noise may make it uncertain when you give "microphone to you". Likewise on CW. Many Europeans use "microphone to you" or "Cambio" by Latin folks.

71. To more accurately determine bearing, swing your beam and watch the S-Meter, this will better ascertain if that P5 is where they should be.

72. On weak DX contacts instead of repeating 4 by 3 repeatedly and the DX station is not getting it, count the numbers out, 1,2,3,4 pause 1,2,3.

73. Some operators operate successfully with no AGC in the CW mode using RF Gain and AF Gain riding. This may work for you in SSB for very weak signals with no big signals around. Also another operating technique is when experiencing heavy QRN is to try no AGC and widen the filters so they respond faster with less ringing

74. During a ragchew, engage the AGC control to prevent crashes and noises from coming in between words of the incoming station. Slow AGC is useful for maintaining constant audio output in roundtable conversations with stations of different signal strengths. For DXing, use fast AGC and use your RF Gain control. Using fast AGC prevents the DX cops from blocking your receiver for the length of the AGC delay. Recommended practice is to adjust RF gain until the AGC is inoperative and set the audio as required. When a strong signal appears, as is quite common in a DXing or contesting, the AGC saves the eardrums and gives time to readjust the controls, if necessary.

75. On CW, try using the SSB filters instead of the CW filters. One noted contester uses 1.8 KHz in conjunction with a 2.7 kHz filter with a touch of CW VBT. He uses the 500 HZ CW filter very rarely. Also use CW REVERSE in a split DX operation, minimizes the pileup racket

76. There is a curious psycho-acoustical phenomena discussed in scientific articles. A weak signal that is "almost free of noise" seems more difficult to hear than one surrounded with a reasonable amount of broad band (white/pink etc.) noise. The noise seems to provide something the brain needs to decipher the signal. Thus, using the SSB filter in a CW contest may result in the ability to copy signals that are difficult to copy using a narrow CW filter. Also some operators have noted that for weak signal detection, try using a wide open filter, seems to bring out the weak signals that are not copyable with the narrower filters. Your mileage may vary.

77. There is another filter that is not in your radio, it is in the noggin. The human brain is unique in that it can concentrate on a particular sound or voice to the minimization of others, i.e., tuning out your Aunt Sally while locking in on another interesting QSO. This works well for some CW ops hearing two or more CW signals at a different pitch. Try practicing this on CW but again your mileage may vary.

78. Some operators use stereo phones with a variety of filters in each earpiece to create a pseudo stereo effect. Experimentation is in order. Reverse the wiring on one earphone to have out of phase audio, seems to work in copying some signals.

79. Be prepared at a moment's notice to go QRP when the DX asks "Any QRP??" With modern rigs – you are a QRP station (5 Watts or Less) – just kick off the linear and run the RF power control to min. The DX might ask to hear you at 100 Watts, so don't cheat.

80. Want to chat with DX instead of ur 59, bye?? Learn something about the country and its history. Ask history questions of the British and they will talk your ear off.

7-4. WORKING THE PILEUPS

81. In a sustained pileup, the callees key up immediately after the QRZ and the strong survive. Others wait a few seconds for the pileup to subside and then transmit their call. I say to myself "Whiskey Six American Dog" then scream away. This works well when the DX operator can't seem to pick out anyone on the initial surge of calls.

82. Know how to operate split using your remote VFOs and TF Set controls as well as RIT and XIT. Some of the interference with a DX station running a pileup is probably due to those who accidentally forget to push the "split" button on their VFO's. YAESU radios are particularly prone to this problem. For example, setting up split operation on a FT990 is a tedious 4-step process and requires extra care to prevent accidentally transmitting on the DX frequency. The momentary T/F button on Kenwood radios require that you continuously hold the button while listening on your transmit VFO, but as soon as you release the T/F set button, the radio reverts to the receive VFO and it is rare that one transmits on the DX frequency.

83. Hang in the pileup no matter what. Can't tell you how many times a European is working the east coast and queries "Any west coast??" Or vice-versa Asia to the east coast.

84. If a station is calling for a geographical area only (e.g., QRZ Europe only), wait it out. If the op slips and just sez QRZ -- pounce on them (they goofed). She/he didn't say Europe in that call, so fair game. If they question you – just say sorry I heard you call just QRZ. Works more times than not.

85. In a "simplex" pileup, always have your remote VFOs ready for split operation, when simplex is not viable, the DX suddenly announces "listening 200 to 210" and there you are fiddling with A=B, VFO 2 transmit, etc

86. You are in six land, and the DX is working by call area, having just finished sixes when you tune in. Put the frequency in memory (Quick Memo on Kenwoods) and tune elsewhere checking the progress of districts periodically.

87. The DX is working by district and you're a W4 living in California, strange how many can be portable in this situation! Usually the DX will specify "Stations in the W# district only". Otherwise your decision.

88. When a DX station gets fed up with the pileup and announces QSY but not where – try going to the bottom of the sub band – more times than not, folks start at the lowest frequency and listen upwards for a clear frequency.

89. Another trick for working a non-split pileup is to say your call quickly without phonetics.

90. When the DX announces "going to ___ meters", Be prepared to change bands in a moment's notice, have your linear and tuners marked for the bands, you'll beat the mob this way. As mentioned before if it is a DXpedition who has established op frequencies, have these in memory, both SSB and CW. Be ready!

91. Often DX stations will call for West Coast only or East Coast only. With the U.S.A. as large as it is, this gets awkward at best. Perhaps you and the DX think that 6's and 7's are the west coast, but what about Wyoming, Colorado, or Arizona – are these West Coast?? Or is the East coast just 1's, 2's, 3's or maybe 4's? The point is if you are waiting for DX to say Midwest Only - you will probably have a long long wait. You be the judge as to what is fair and prudent. Until the DX learns to say Midwest only or West of the Mississippi – well you get the idea I'm sure.

92. On occasion a DX pileup will get a QRX, everyone standby, but in reality the DX station is listening for something they need or listening for the Little Pistols. Then Son Of A Gun, he makes his own "list", comes back and calls off his list. You can take it from there.

93. In a split operation, the DX announces listening 200 to 210, Yeah sure as he picks up a weakie in the clear at 212.

94. The DX is working stations outside the USA and you hear a rare African station calling the DX. Throw in "there is a African station calling you". DX asks - who had the African info? That's you and you just made the log and he/she just snagged a TY.

95. Use your TF SET control or VFO A & B controls to figure out if there is a pattern to how the DX station is picking out the callers. See Chapter 5, working split. Maybe you want to avoid calling on a frequency where a Big Gun is parked, then again the Big Gun is likely to get through the pileup well before others and you might want to tail-end there or be up/down 2 or 3 kHz ready to call when the Big Gun works em.

96. When you are working DX, pileup or not, and the cadence goes on and on with no QSL information --- don't be afraid to ask QSL INFO PLEASE. Most DX operators will give QSL info periodically. But if not ask, otherwise you will be searching the Internet sources for hours and bugging the DX Reflectors with "Anyone have QSL info for DX#DX??"

97. QSL via burro means send your QSL via the International QSL Bureaus (for USA via the ARRL). QSL via CBA means Call Book Address - send direct to the DX Station direct using the address listed in a callbook (RAC, Buckmaster, QRZ, etc.) QSL via HC means home call, send cards to his home station, either via the burro or direct. This is typical for portable stations AC6V/ZD9 – don't I wish! Also I have heard QSLL meaning the DX will send his card when he receives yours.

98. Listen to DX QSO's if you have nothing better to do. DX asks "anyone know how to reset a TS-850?" You do and you are in like Flynn.

99. You are a little gun and the pileup is enormous, no chance you say and the DX asks "Any one near Podunk, Iowa can get W0xxx on frequency?" Are you in Iowa?? Bottom-line listen to the pileup under any circumstance, opportunities do arise. Especially if you have a high country count and this DX may not appear for several years.

100. Don't be a DX pig. If you have already worked the DX station, give someone else a chance. Keeping someone from working a new one so you can tell the DX you got his QSL card – is a sure way to get labeled as a DX Pig.

101. Chit-chat with a DX station during pileups is not a good idea; the 300 other folks on frequency won't be pleased. But if the DX station wants to or knows you – no problemo, they are the boss (or should be). During a massive pileup, asking what antenna the DX has is dumb at best.

102. Asking the DX station to QSY to different band is also a no-no in a big pileup. But if its your last country for 5Band DXCC or 5Band WAZ – you decide.

103. Don't ask the DX station in the middle of the pileup, to work your buddy with an attic antenna! This holds up the pileup and the DX station is already listening up for him and everyone else.

104. Don't try to help a station. If the DX is having trouble copying a station, let them work it out, I for one resent cop assistance.

105. After a phone contact, asking the DX to switch to CW for a quick contact is done all the time – after all they can say no or make an excuse. Yes - CW is allowed on the phone bands.

106. In a pileup, don't ask the DX station to check to see if you are in the log from a previous contact on another band or mode. Thousands will hate you!

107. A DX station states he doesn't QSL – doesn't have any cards. Using your computer, print one out as if it were his/her QSL card - all filled in so all they have to do is sign it. Best bet is to send an SASE with his country's postage - send it off and pray. Worked for me.

108. No cards are coming out of a certain country although lots have worked em. Have a friend write the address in their native language. Much more on this in QSL Tips.

109. Some Dxpeditioners – actually call CQ away from the published Dxpedition frequency – better know when they usually come on and Tune, Listen. Tune, Listen. This is a favorite strategy of QRPers catching that rare Dxpedition. Also why not call the DX station 5 minutes before they usually come on the published Dxpedition frequency??

110. Some DXpeditions use "trickery" in controling/working the pile up.

    Call: . . . .   "This is DX11DX listening up 5"
    Next Call ... "This is DX11DX listening up 10"
    Next Call ... "This is DX11DX listening up 25"

    DX11DX then listens up on 25 only. The DX station has managed to thin the pile up out since only the stations listening for awhile are listening up 25, the rest are down below. If you're the "callee" then the message is to "listen listen listen."

110A:

    Another devious technique. "The YL only" (when there are no YLs on frequency) OR "The mobile station only" (no mobile on the frequency). This quiets the pile up down and is much better than giving a letter or number or prefix or suffix that is likely to be included, all or partially, by many of the calls in the pile-up. The DX station listens to the thinned out pile up since all are waiting for the YL. Those who are aware of this trickery can jump in.

110B:

    Yet another devious technique is where the DX station transmits on 14.195 and announces, "Listening 5 to 20 up (mob starts calling) (pause) and 14.190." Son of a gun --worked him with one call on 14.190. Listen, listen, listen.

DX DX DX - that's all you ever think about!

🙂 THE OLD DXER

The sunspots were roaring and the Old DXer was working DX night and day.
The faithful XYL had had it with his QRZing and blew her stack:

"DX DX DX - that's all you ever think about!", she groaned, *"Why I'll bet you don't even remember our wedding date!!"*

"I mostly certainly do" was his immediate reply. *"It was June 14th, 1958 - That's the night I worked THE XT1, THE CRØ, AND THE AC6!!"*

7-5. DXing ON CW

111. If you are new to CW DXing, you may want to read Appendix A1, CW operating procedures before continuing.

112. Do get your code speed up. Some DXpeditions only operate CW and the pileups are thinner than phone. I have at least 10 countries on CW that I haven't been able to work on phone.

113. Get some CW pileup tapes – and practice the protocol. Several are available on the Internet such as PED, Pile Up, HSTT, RUFZ and others. See operating aids, Chapter 3.

114. For CW, the beginner can get shaky and even send their call pretty raggedly. Get a memory keyer until you get used to it all. Some of the more exotic ones can be used to memorize your call, report, increment serial numbers, etc. Just push buttons!

115. For CW, listen to determine the pattern of the end of the exchange, just TU. Other times QRZ? or QRZ de X6XXX, or just X6XXX. Some stations send dit dit.

116. For CW Simplex contacts, use the zero beat technique to get right on frequency. Be sure to zero beat with the main tuning dial with RIT off. Some of the newer rigs have auto zero beat, so check the manual.

117. Experiment with your RIT and XIT controls, tune the DX to the most comfortable pitch using RIT, then switch from RIT to XIT, to stay on frequency or work split.

118. For CW, the CW Reverse control will allow your bandpass to "listen" down instead of up thereby avoiding most of the +1kHz to +2kHz pileup racket. This doesn't affect your transmit frequency. Crystal filters tend to ring and drive me to distraction, by using CW reverse; many times works better than filters. Your ears may vary.

119. For CW, learn to exercise the powers of the human ear to lock in on a certain pitch and mentally reject others. Tune around to CW signals with QRM and practice a bit.

120. For CW, try different pitches besides the factory set 800 Hz. Many of the best CW operators use a pitch of 400-500 Hz. There is an apparent noise reduction as noise peaks caused by narrow filters may seem not so sharp.

121. For CW contests – jump in – with computer practice, you should be able to recognize your call at any speed – so you'll know when they come back to you. Just reply TU ur 5NN and the report (zone, etc.). Same with non-contest pileup DX just 5NN.

122. Send 5NN reports in a rapid fire DXpedition or contest. Operators don't want to write down 439, 559, etc., just a line from top to bottom of 5NN. Many have their computers set to automatically enter 5NN. Goofy but a fact of DXing and Contesting

123. For CW, have a set of Q-signals handy, some old time ops have a ton of them in their vocabulary

124. Shoveling – a sneaky technique where the DX announces QRZ up 5 and while the multitude is calling (and can't hear), the DX sends 23 meaning he/she is listening at 14.023 KHz (or the band in question)

7-6. DXING BY CONTESTING

125. Learn how to work the contests. You don't have to be "in" the contest to jump in. Some very rare countries come on during contests but not otherwise heard often. Example SO2 Western Sahara, and ZC4.

126. Get some CW Contest pileup tapes – and practice the protocol. See Chapter 3. Operating Aids. With your computer, have it send your call repeatedly gradually increasing speed up to 30 wpm until your call is a musical sound not dits and dahs.

127. For CW Contests, you should be able to recognize your call at any speed – so you'll know when they come back to you. Just reply TU ur 5NN and the report (zone, etc.). Or push buttons on your memory keyer.

128. Just before the big DX contests, many of the DX contesters flex their muscles by checking things out. This is a great time to catch DX without the contest pileups. In addition, prior to the contest, contesters will operate on the WARC bands, which are banned for contest activity. Also some of the big contest stations often stay on locations for a few days after contests.

129. Near the end of a contest, the common ones e.g., P40 will be begging for contacts. Just learn the report exchange – it is on the web in the rules of the contest calendars.

130. Tune the entire band, you may find a TF at 28.700 during a contest and a ZD8 at 21.400!

131. During a contest, check the other modes of operation, e.g., during a SSB contest, check RTTY, PSK31, digital, CW, etc. Some rare DX can be found this way. Non-contesters will work a different mode to avoid the howling mob.

132. In a contest, you hear a weak CT3 from Madeira calling CQ contest with no takers, but with the 20 over S9 signals all around and QRM galore, you figure no use. Wrong – he is running a dipole with 50 Watts and comes right back.

133. Get a rare prefix and work the WPX contest – DX will call you!

134. There is a big pileup of JA's calling you in the WPX contest, tell them to stand by. Ask for Asia outside of Japan – oopps there is a JT in Mongolia! Son of a gun!

135. Same applies if you have a rare prefix and the USA is piled up on you in the WPX contest (fun huh). Stop the USA guys and ask for Africa or the Middle East. Son of a gun -- a 4X from Israel calls you!

136. In the WPX contest, if you are the callor instead of the callee, you want to learn how to run a pileup. You are in charge; make it very clear "How things work". No tail ending, no calling out of district turn, I said Asia Only, etc. Announce your call frequently and any QSL info. If you just do QRZed, a bunch of "what's your call" will drive you goofy. If the last thing you send is your call, the other operator is more likely to remember it.

137. It's OK to ask for QSL info in contests, if they don't give it after a long time. After all he/she wants your points and you want their QSL. Most contesters will QSL.

7-7. DXING BY OTHER MEANS

138. For shaky contacts, work "insurance contacts", preferable on a different band or mode. Amazing how many times the DX logbooks show 3 contacts where I'm sure I had four on different bands, modes. Very important as some DXpeditions come on once every 20 years, you want to be positive you are in the logs.

139. Someone throws a carrier on the DX – no problem, the NOTCH control will minimize it and the new DSP Radios will all but make it disappear.

140. Operate RTTY - many rare DXpeditions and countries are operating in this mode. Also PSK31 is becoming more common. Don't overlook 10 meter FM.

141. Learn some common signoffs with DX stations. Japanese (Sayonara), Spanish (adios), Russian (Do svidanja), French (au revoir), Aussie (G'Day), Italian (arrivederci, Ciao). Also other phrases such as: hello, how are you?, welcome, goodbye, please, thank you, what is your name?, my name is..., do you speak English?, yes, and no. May endear you to them when comes time to QSL. See URL: http://www.elite.net/~runner/jennifers/goodbye.htm

142. On nets, learn the jargon, last two is the last two letters of your call, nickels – nickels is a 5 by 5 signal report. Last heard means you didn't really hear the DX station, but you are mimicking some else's perception – just kiddin.

143. A watery, obviously DX carrier comes on, check it out, could be rare DX.

144. Listen Listen Listen

7-8. EQUIPMENT

145. For your linear or older tune type transceivers, make a card indicating where the controls are set for each band – saves time and in these days of DX Packet Clusters – time is of the essence.

146. If you have a low-pitched voice, equalize your audio. Ever notice how the YL's and kids get thru the pileup. High-pitched tones seem to get through. Don't use your local non-DXing buddy to evaluate this. Try several weak Europeans for reports.

147. On phone, pack in your audio, use speech processing, and equalization, but don't splatter. Have a local DXer check you audio levels. This point cannot be over-emphasized; the bands are loaded with under modulated signals.

148. When having others help adjust your mike gain and speech processing, do not have a non-DXer listen to your adjustments, many rag-chew types wouldn't know good full modulation if they heard it, they go by the ALC settings which may be way too conservative.

149. When adjusting audio, if possible, swap operating QTH's with a DX buddy. That way you can hear your rig on the air and evaluate the settings. Borrow a bandscope if possible or a high-frequency oscilloscope.

150. Don't necessarily rely on your ALC for the correct audio modulation, manufacturers tend to be conservative, use a band scope or a buddy to determine the max level without splattering. You can tune the bands and hear dozens of under modulated SSB signals.

151. When working weak SSB DX or CW, turn the RF Gain down and the AF Gain up, then RF Gain up, til you can hear the DX. Helps with apparent signal to noise.

152. Try using the IF Shift control as a quasi-tone control. It can be adjusted to cut high or low tones and might give a better audio response to copy that weak rumbly or noisy DX station.

153. Tune slightly above (on USB) the DX Frequency, this will put more high tones in your signal to the DX. Tune slightly lower on LSB. Good place to use XIT – See Chapter 2.

154. If you have a vertical or live in a noisy area, get a radio with DSP, or an outboard unit, they do work – sort of, and they are getting better.

155. Got a windfall – maybe buy a new rig? Nope – put it into the antenna. I'd rather have a Swan 350 with stacked monobanders, than a TS-870 with a vertical.

156. Got a windfall and a gang buster antenna – try an 800 watt linear -- will get you 9dB gain (1 ½ S-Units) which can get your signal up out of the noise for those weak contacts. Less than 3dB more when increasing from 800 to 1500 Watts. This will also make you competitive on the lower bands where noise is a killer.

157. Set your AGC to fast since AGC decreases receiver sensitivity but you want max sensitivity for working weak DX. With slow AGC, QRM lowers the receiver sensitivity and the long AGC time constant increases the time that weak signals are down in the mud. Use the RF gain for optimum copy.

158. For headphones, long sessions will require a comfortable set. Depending on your age, high fidelity phones may be better for you than "communication phones". For me, the high notes seem to be far more important for intelligibility than the bass notes, and a pair of Radio Shack Titaniums are my favorites. Some operators use the stereo phones for monitoring different receivers and some use a variety of filters in each earpiece to create a pseudo stereo effect. If you use hi-fi phones and the excessive high frequencies bother you – some filters will tailor it to your listening preference.

159. Speakers. Save your matching $120 speaker for the strong signal rag chewing. A headset is the only way to dig out the weak ones in heavy QRM. Shuts out XYL (OM) and harmonic noise as well.

160. If you are experiencing splatter from an adjacent frequency, be sure your Noise Blanker is OFF, some rigs allow blowby of strong stations in this mode.

161. Consider your backup equipment. Even a Swan 350 will get you on the air for that rare weekend DXpedition if the main rig goes QRT. Spare coax is a good idea, I recall blowing out a length of it while tuning a linear – in a clear spot of course!

162. Got noise, try turning the beam, beam widths are wide, contrary to popular belief, and you can minimize noise and still get through.

163. GROUND GROUND everything. A friend of mine rented a hydraulic drill and went down deep for his ground and his noise level decreased significantly.

7-9.   OPERATING AIDS

164. Study propagation. For propagation predictions See Chapter 4, but don't be surprised if you can work an S79 when Miniprop says no way.

165. Use the NCDXF/IARU Network to determine band openings, programs are available – see Chapter 3.

166. If you are not totally familiar with prefixes, have a desk card available to quickly determine the country – you may need em. Example an HM1DX comes on the air, better work em cause that's North Korea (P5), but probably not legit!!

167. Always have a want list handy and coupled with above, concentrate on when and on what band a new one might be found. For example at the waning days of the last sunspot cycle, I had worked very few Europeans on 12 meters and noticed a very narrow opening of only 15 or 20 minutes in the morning. A German station told me I was the first W6 he had worked in months on 12 – sed I called at the right time, yep but I knew that.

168. Great circle bearings, beam headings. Either on your computer or by a hard copy, you will need beam headings from your QTH to the DX station, unless you have a vertical. Check the Internet for freebees -- DX View 1.1.5 finds heading, country, distance, and zone info from a callsign prefix entry. For a small shareware fee, WinHdg -- By WA6FHI finds similar info and sits in the background on your computer until you need it.

169. If you got all the common ones, use DX View, with this you can get alarms for the needed countries, your computer will squawk at you when it gets DX spots from the internet that match your "alarm countries" URL http://www.dxlabsuite.com/ some DX Packet cluster software will do this also.

170. If you have language difficulties on phone or CW, use Q Signals as much as possible. There are Q-Signals for many situations that experienced CW ops will know.

171. Listen Listen Listen – where? There are informal, defacto DX frequencies and windows – know what these are and check regularly. See Chapter 5.

172. In addition to the Chapter 5, Finding and working DX, and Chapter 9, QSLing, here is a quick list of DX Tips gathered from many DXers over the years

173. You have heard it before, but it's on nearly everyone's number 1 rule for DXing. Listen, listen, tune, tune, tune, if you can catch the DX before the pileup begins or the Packet Clusters make it known world wide, you can find a lot of DX calling CQ.

174. Learn to distinguish a DX signal by its characteristics, polar flutter, hollow sounds and fading unlike domestic signals. For SSB, the operator's accents and pronunciations can be a tipoff. Drifting or chirping CW signals are another clue indicating poor power regulation in some countries.

175. When calling in a non-split pileup, try delaying your call for as long as it takes to say a 5-letter call. Works sometimes for split operations also.

176. Another trick for working a non-split pileup is to say your call quickly without phonetics.

177. Some callers give their last two or three letters only. This works for DX operators who can't seem to get a complete call. You'll hear " the WD6 – go ahead", this is a tipoff how the DX op is perceiving the pileup. Using incomplete callsigns is a questionable practice, as the rules require a full callsign. Also it slows down the Q-rate, taking longer to complete the exchange. Most good DX operators discourage the practice.

178. When a DX station suddenly disappears, don't give up. Call periodically, they may and sometimes do come back. Nature does call you know!

179. Call the DX by name after your call sign if you know it. DX ops get tired of the routine sometimes. If he asked where you met – Dayton is as good as any.

180. If you are at the far end of the feeding frenzy, give your location. DX stations love to hear the West Coast when the W1's are bombing in. And since Wyoming or Rhode Island have relatively few operators; throwing in your state can't hurt.

181. Don't give up when the DX goes back time after time to a one region (East or West Coast). All of a sudden you may hear any "West Coast W6 or W7". This has happened countless times. "Any YL's" is another – no squeaking male voices please.

182. Be ready to turn your power down when the DX calls for QRP stations. Remember 5-Watts folks. No fair calling with a kilowatt, then turning to QRP levels.

183. Where is the signal coming from? Great circle you say, well don't bet on it. There have been numerous cases of skewed propagation paths to make it worthwhile to swing the beam if you have one. Verticals don't care, of course.

184. Short path or long path? Swing the beam and see. Long path signals can have a hollow effect to them. And of course the time of day is another indicator. West coast propagation to Africa is typically long path on 20 meters in the morning.

185. Listen with a vertical or dipole and transmit with a beam, not that a beam has that sharp a beam width or front to back attenuation, but you may be dealing with weak signals, and omni directional antennas may make a difference.

186. On the low bands (160 and 80 meters), a low-noise receiving antenna is almost a must. Beverages or loops are often used.

187. Tune the entire band – ZD8Z was once heard a smidgen away from HCJB!! 21.450 MHz that is. On 10 meters, Iceland was once heard at 28.700 MHz!!

188. When finding a DX station rag chewing, don't break in. Put the frequency in quick memo and check back often. "I'll take a few for a new country" the DX says after a ragchew. You may assume this means a new band country – if not specified otherwise – hi hi.

189. When the DX is working by district in numerical order, honor the DX's wishes. If they just finished your district and you are in for a long wait, store the frequency in quick memo and check back now and then as you tune about to initiate a new pileup by catching em first.

190. The following items are about nets, lists, skeds and packet clusters all of which are highly controversial among DXers. So without taking sides, the author wants to say this about that. Early on in your DX endeavors or even now you need to make some hard decisions.

191. Work the nets? – it's your rig, your call, your choice, but I wouldn't mention it at the DX Club. Many DXers hate the nets and list operations – be aware, if you are peer conscious. To be clear – the author doesn't approve or disapprove – I just don't care.

192. Make a sked with that needed country? Well it's a lot like a private list wouldn't you say. Your call.

193. Use the DX Packet Cluster for getting new DX and never make a spot yourself? Hmm. But at least you still have to break the pile, unlike the nets, where you only have to get to net control, regardless if you can hear the DX or not!

194. Borrow a buddy's big station to get the last country? Contesters do, whatta you think??

195. Don't be a band junkie. Check all bands for DX. Africa can come in on several bands during the day.

CHAPTER VIII   CONTESTING

8-1. CHAPTER CONTENTS

This chapter contains information on contesting. The table below lists the chapter topics and paragraph numbers where they can be found.

NOTE: Throughout this section, reference is made to the Callor – that is the one running the pileup. The callee is the poor guy trying to get a contact.

8-2. THE ART OF CONTESTING

Contesting is a grueling dog-eat-dog competition that requires a fairly husky station, stamina, and dedication to be the best you can be. You can participate in a contest whether or not you decide to formally submit an application and your logs. There are easily over 200 contests per year - four contests per week! Besides magazines and publications, one of the best sources of contest activity is LA9HW, Jan Almedal's Contest Calendar Pages at URL:
http://home.online.no/~janalme/hammain.html

Jan's pages lists contests by month and in most cases includes the rules for report exchanges, application, and classes of stations in the contest. Some contests have categories for 100 Watt and QRP stations, but a good antenna is a must if you intend to sit on a frequency and invite stations to call you. As the caller, you can roam the bands and work as many entities (DX, Prefix, States, etc.) as you can. One of the best ways to get into contesting is to be the caller for a few contests and roam the bands and do a search and work as many QSO's and multipliers as you can. From this you will soon learn the rhythm and protocol of contests.

When you get proficient as the callee, you can try a casual approach to being the callor where you sit on a frequency and let folks come to you. Next is finding the local contest club, field-day group, or big gun contesters and participate as a full-fledged contester. With these groups, you can learn the contest secrets of the ages, as these folks have been at it for years and know all the tricks and techniques.

Note that during a contest, unfamiliar reports are given 592345 is a serial number, 5906 is a zone – see rules

There are a wide variety of categories in contests, following are typical, but check the contest rules for entry into any contest:

Single Band or All Bands

Single Operator Unassisted -- One person performs all operating and logging functions. Use of spotting nets (operating arrangements involving assistance through DX-alerting nets, Packet Clusters, etc.) is not permitted. Single-operator stations are allowed only one transmitted signal at any given time. (Note: This does not permit multiple single-band entries from the same station).

Single Operator Assisted -- One person performs all operating, monitoring and logging functions. The use of spotting nets and assistance through other alerting systems not physically located at the station (operating arrangements involving assistance through DX-alerting nets, PacketClusters, etc.) are allowed.

Some contests such as the WPX have a TS category, i.e., Single Tribander or Single Element 160/80/40/

Multi-operator -- More than one person operates, checks for duplicates, keeps the log, etc.

Single transmitter or multi-transmitters.

Power categories; QRP - 5 watts output or less, Low Power -150 watts output or less, High Power - more than 150 watts output.

So there is room for the Little Pistol. (Best have a good antenna though.)

From the ARRL General Contest Rules - "3.14. In contests where spotting nets are permissible, spotting your own station or requesting another station to spot you is not permitted." And from the Rules for Contests Below 30 MHz, Single-Operator Category - "2.1.1. Use of spotting assistance or nets (operating arrangements involving other individuals, DX-alerting nets, packet, Internet, etc) is not permitted."

From the CQ WPX 2002 rules - "Passive use of packet or internet DX spotting nets is permitted for [SOA, MS, and MM] stations only. No self-spotting by a station or one of its operators is permitted." From the section on Disqualification - "The use by an entrant of any non-amateur means such as telephones, telegrams, internet, or the use of packet to SOLICIT contacts during the contest is unsportsmanlike and the entry is subject to disqualification."

From the CQ WW 2001 rules on Single-Operator category - "The use of DX alerting assistance of any kind places the station in the Single Operator Assisted category." A similar statement to WPX rules is found in the Disqualification section.

8-3.   PHONE CONTESTING EQUIPMENT

For the phone contests, it is highly recommended you have a computer (low EMI type), a Digital Recording Unit or Digital Voice Keyer, headphones with a boom mic, and a foot switch. Computers offer a multitude of features including logging, propagation predictions, packet networks, radio and rotor control, automatic CQing on CW or voice, automatic report sending, and much more. The computer programs will save you lots of time and grief by providing you with such aids as duplication checks, Q-Rates, cw speed, entities, zones, prefixes worked, data lookup and automatic tallying of totals. For programs, see Chapter 3, DXing/Contesting Aids.

When evaluating a contest program, be aware of the variables that might occur. With all the options and features, they can react differently depending on the contest.

Some features change based on entry class, type of hardware connected, and possibly the band or mode selected. Since it is not possible to test these programs in all configurations, test your specific configuration and make sure everything is functioning properly.

A DRU (Digital Recording Unit) or DVK (Digital Voice Keyer) is a device that records your voice and calls CQ for you at the touch of a button. These are available as an option for internal installation on most modern rigs or as an outboard unit. The DRU accessory is a must as during the contest you may have called "CQ Contest" a several thousand times. During the last hours of the contest many a contester has ended up hoarse from the repeated calling.

The foot switch and headset with a boom mic allows for freeing both hands for computer typing or manual logging if you insist. Using VOX is an art and you will find the delay and gain settings to be critical, best use a foot switch to key the transmitter. If finding entities (prefixes, zones, etc.) is taking too much time by computer, some operators use a check sheet on the desk.

8-4.   CW CONTESTING EQUIPMENT

Again a low EMI computer is the best way to log and keep track of your performance. A memory keyer is essential unless you want to send "CQ Contest" several thousand times by hand. And keep in mind that the callors operate at very high speeds. With many operators, the memory keyer speed is much faster than the operator could send manually. Contesters usually operate most of the contest at very high speeds, only slowing down when no one answers, so as to entice the non-contester types and the slower speed operators.

This is typical in the waning hours of a contest. The memory keyer also has a host of features that assist in a contest, such as automatic increment of serial numbers for those contests that require them.

Many contesters operate QSK so they can hear between sending and this may require equipment mods such as a reed relay – vacuum relays in the transmitter (linear amp) circuits.

8-5. PHONE CONTESTING

Brevity is the name of the game in a contest. Short calls repeated frequently is much more effective than the standard method of 2x2 CQing. Contest Contest, AC6V; CQ Contest CQ Contest Alpha Charlie Six Victor, are favorites. But be careful with the cuties, DX operators may not understand your jargon. Pause just long enough for a response, then repeat your calling. If the last thing you send is your callsign, then the other operator is more likely to remember it.

Contesting is a quick paced affair and waiting too long between calls is lowering your Q-Rate and inviting a big gun to move onto your frequency. Don't bother to ask "is the frequency busy". Pick a clear spot and sit on it to establish dominance. Give 59 (contest peculiar) reports only. If you are not getting responses, ask someone if your frequency is clear.

A big gun (you can't hear) may be wiping you out. Call about every two seconds or so, repeatedly, folks are zipping across the band so let them know you are there.

If several stations respond to your CQ and you don't get a complete callsign – transmit "the station ending xyz go ahead" (or the part of the callsign you did get e.g., the kilowatt four – go ahead.) If you didn't get any of the call – simply transmit "again" or QRZ".

8-6. CW CONTESTING

Before jumping into the contest, in addition to tuning up everything to a razor's edge, load up the memory keyer with all messages you will need. Amazing enough in my conversations with well-known contesters, some send at speeds twice what they can copy!! So send fast code, it increases your responses and jacks up the Q-rate. Remember the exchange is short, expected and just about anyone can recognize their callsign at any speed as well as the report. Slower copiers may listen to your call 2 or 3 times to get it correctly, so most contesters send very fast, not to show-off (how can you show off a memory keyer?) but to increase Q-rate.

If you are a slow CW operator or just starting -- here is the secret for getting into a CW contest. First practice on your computer, send your call to yourself repeatedly, gradually increasing speed. You will get to the point where you can copy your call at any speed – so when the contest station comes back to you – you will know he/she is calling you.

Next load up your rig's CW memory with your call. And in a separate memory, load up with the report (some memory keyers will increment serial numbers for you). Always use 599 as the RST. All this can be done with a Contest program as well.

Now you listen to a station calling CQ contest at high speed – copy as many letters as u can, keep listening until u have his entire call. And keep listening until you have the serial number (or zone, etc) if used.

When they call CQ or QRZ – hit the memory button that sends your call. If they come back to you – hit the memory button with the report. If they send IMI (question mark) just hit the button again. THAT IS IT. Notice you did not send the call of the contest station, only your call and if they come back – the report.

Some contesters are operating at 40+ wpm to save precious seconds in this highly competitive environment. Asking local contesters here in San Diego revealed that several send at about 25 to 30 wpm, slowing down a little when there are no big gun takers. Occasionally if no one is calling, reduce the keying speed to entice the slower speed operators.

Find a clear frequency if possible, send your CQ contest call, don't bother to send QRL (busy) and waste time. Dueling for a frequency spot during a contest is just part of the drill, so pick one and fight it out. Call, then wait for about two seconds or so, and repeat. Potential callers are zipping across the band so let them know you are there. Call CQ TEST AC6V

When you get a rare country calling you, ask them to QSY for other band contacts.
Just because they are rare doesn't mean they are calling CQ, they may be in the contest casually or can't handle a CW pileup. If a caller is too fast for you – just reply "AGN" or "?" If the high-speed caller persists, maybe a "sri, no cpy" will suffice.

If your q-rate drops below a satisfactory level, maybe below 10 contacts an hour, go on a search and pounce routine. Pause during a lunch break and quickly determine the multipliers you need, your computer can assist in this, then start tuning to fill them in. Tune the bands for a new country, zone, or multiplier.

There may be only one Western Sahara on the band or one zone 40. You can increase your score significantly by doing this. If you can't copy the callor's call, work them anyway, then listen for 2 or 3 rounds copying a letter or two each time.

Assisted or unassisted is a big question, as many contests require you to specify whether you are on a local net or packet cluster. By using the packet cluster, it can net you lots of new countries, zones, etc., if you are impatient with search and pounce. In contests where packet or spotting nets are allowed, spotting your own station or requesting another station to spot you is usually not permitted.

The top contesters are aware of propagation and the band openings to the various parts of the world.

Study these before the contest to know what areas to concentrate on to maximize the Q-rate. Die-hards check propagation 7 and 28 days before the contest. Since solar conditions tend to repeat every 28 days, operating four weeks before the contest can somewhat forecast solar conditions on the contest days.

Operating several days before the contest will help you find the likely long-path and gray line openings, even though solar conditions may be different.

Study your contest results from previous years and set goals based on last year's performance. Compare notes with other contesters to see where you might improve your strategies.

One of the hardest things to learn in contesting is when to stay with a run frequency that is not really productive, versus going off for search and pounce. In the medium to little pistol category, use a Q-rate of 60-90 per hour as the break point. If your rate is 60 or better, you should probably stay where you are. A rate of 60 per hour is one QSO per minute. That seems pretty slow, as you will probably call CQ four times for every answer.

Getting to the band early makes it much easier to find a clear frequency to get started on. If you're planning to go from 20 meters to 15 meters at the band opening, you need to monitor 15 on a second radio, and go as soon as you start to hear signals. Find your frequency and start making QSOs.

Understand that when you are already running on some frequency, others will have to choose how close they want to get to you.
Your mere presence on that frequency should make them understand that getting close is going to be a problem for them. Some tips for holding a frequency:

1. Transmit frequently! Lets interlopers know you are using the frequency.

2. Transmit a clean, strong signal. Don't pump up the processor too high.

3. Don't necessarily optimize your transmit antennas for F/S and F/B. Let your competitors hear you and hopefully avoid the frequency. Also you want to be able to hear the DX calling from different areas of the world.

4. If others move in on you, let them know you can copy through anything. Tweaking on DSP, Notch, IF shift and IF width can make a big difference

5. Use directional receive antennas optimized for the key directions and vertical angles with plenty of F/B and F/S. Try beverage antennas on the low bands

A rate of 60/hr for 24 of the 48 hours is over 1400 QSOs - a very respectable total! Looking at the statistics, based on minutes per QSO. 95% of QSOs will be made between .5 and 1.5 times the basic rate. This means that you should be able to tell what your rate will be after just a few QSOs. So don't waste more than about 5-10 minutes trying to establish a run. If after 5 minutes you haven't worked at least 3 stations, try a new frequency or a new band. Run frequencies are not really that hard to find.

It is very common to send RST reports in abbreviated form during a contest. These are called cut numbers or abbreviated numbers. For example 599, is sent as 5NN. "N" in place of the number "9". Also another time saver is for the zero using a long "T". "T" is sent in place of the number zero as in "POWER HR IS 3TT WATTS".

There is a number code for all numbers; however, the N and T codes are the most common ones. Also CW stations sometimes report their zones as "A4" or "A5" instead of sending "14" or "15". 1 = A, 2 = U, 3 = V, 4 = 4, 5 = E, 6 = 6, 7 = B, 8 = D, 9 = N, 0 = T

8-7. LOGGING, Contest

In ye olden days before computers for hams, logging was done by hand, requiring time or an assistant to handle the task. Duplicate contacts were a big problem, requiring a thorough checking after the contest prior to submitting the logs to the contest manager. With the advent of the ham computer and contest programs, dupes can be avoided real time and has a lot of features that minimize the totaling and preparation for submittal.

Features include automatic incrementing of serial numbers, 50-line VGA display mode, color-coded band map, window position and color control, mouse support, sunrise/sunset tables, band switch support, variable CW spacing, increased CW speed range, Q-rates, beam headings, the continent worked, distance, rotor control, single op/multi TX support, duplicates (including partials), lists multipliers by band, country look-up function, lists all contacts, and provides many current statistics. An important feature is the ability to import and export files from/to different formats.

Programs are available to handle RTTY, CW, Phone, Multi-Ops and Networking, two radio operation and more. For a variety of contest programs – see URL: http://ac6v.com/dxsoft.htm

8-8. SUBMITTING THE LOGS

Be sure to thoroughly read the contest rules for submission criteria. Use only approved contest forms, or a reasonable facsimile. Paper logs are still acceptable but with some contests, entries can be on diskette or uploaded to a BBS. Some contests even allow submissions via the Internet.

Depending on the contest, in general, logs must indicate times in UTC, bands, mode, calls, complete exchange sent, complete exchange received and QSO points. Multipliers should be clearly marked in the log the first time worked. Entries with a high number of QSO's, may require cross-checksheets (dupe sheets). See the contest rules for more at LA9HW's page at URL: http://home.online.no/~janalme/hammain.html Or the Contest sponsor – ARRL, CQ Magazine, etc.

8-9. CHECK LOGS

Most of the contests use some of the submissions as check logs. These are useful to help in tie breaker situations, but probably are not used much beyond breaking ties or verifying questionable entries. Just be sure your entry has all the necessary information as specified by the contest rules and that you have checked for dupes. Some contests use the UBN system to express the percentage of accuracy of the log. U is unique – only one in the contest that worked the station, B is busted call, and N is not-in-the-log.

8-10. CONTEST CHECKLIST

BEFORE THE CONTEST

Make a checklist like this one, customized to the needs of your individual station.

Test everything over and over, until you are satisfied no glitches exist.

Set Computer to exact UTC time.

Set up computer files, cw/voice memories, keyboard overlay. Test DRU (DVR) units for clarity.

Simulate a few QSO's on the computer, with rig interfaced, then erase log file.

Obtain propagation forecast and make Miniprop readouts.

Review past contest logs and magazine results.

Update amplifier tuning chart, tuner memories, and check for rfi at 1500w, all bands, computer interfaced.

Check that computer boots clean, with no unnecessary TSR's.

Verify attenuator, AIP, notch, noise blanker, split frequency all off. Check vox delay.

Use your own key paddle – adjusted for contesting.

Adjust CW pitch for long term listening

Contest rules handy.

Prepare sheet with suggested frequencies. Prepare off-time sheet.

Do a receiver noise survey with computer on. Does turning ant slightly help?

Establish a difficult, but achievable, goal for the contest.

Look at last year's rate sheet to fine-tune strategy for band changes.

Domestic contests - max antenna height 10m-50', 15m-75', 20m-100

In the final hour before the contest, tune all the open bands and spot the rare stations that are testing and holding a frequency for the beginning of the contest. When the contest starts, work the noted stations before the packet cluster hounds spot them.

DURING THE CONTEST

Ignore other peoples' numbers. Some play games and off-times are unknown.

Very short pause between CQ's, so nobody can tune past while the frequency is silent.

BAD - Who was the Yankee Zulu? GOOD - Yankee Zulu 59(rpt) AC6V

Every QSO is important.

No alcohol, except nightcap Sun a.m. and victory celebration Sunday

Population density - Western states advantage - Call CQ. Use the highest band open, to avoid 9's working 1's, with the 9's aimed East.

Know who is running above and below you. Keep centered, but not too far away so somebody can sneak in.

Don't give up the frequency – dominate!

Use automated cw and voicekeyer

DURING THE CONTEST - CW

Start at high speed (32-35), then slow down as the rate falls (26.)

Adjust speed so ONE station is tail-ending.

If a pileup grows, increase speed (40+) until it becomes manageable.

AGC off. Ride the RF gain control.

Tune RIT +/- 400 Hz after CQ.

Go high in the band (050) and send slow (25wpm) cq's once in a while.

In Search & Pounce mode, use LSB (cw reverse), tune from high to low to maximize collisions.

DURING THE CONTEST - SSB

Attract non-contesters with plaintive CQ's. "Anybody, Anywhere" "You don't have to be in the contest"

Pause during a JA run and ask for Asia outside of Japan.

Pause a USA run and ask for EU, AF. AS

During a lull, swing the beam 180 deg and try the other part of the world.

Insist on the full exchange.

Use Fast AGC, only to protect your ears. Ride the RF gain control to avoid compression.

Prolong 10 meter novice operation beyond maximum rate. These are unique contacts, and it will improve your nightime rate, when you work the "real" contesters.

MAJORUS
LUMPUS
COMET

HOUSTON YOU
HAVE A PROBLEM
DOWN THERE

CHAPTER IX   ALL ABOUT QSLing

9-1.   CHAPTER CONTENTS

This chapter contains information on QSL cards and QSLing. The table below lists the chapter topics and paragraph numbers where they can be found.

9-2.   THE QSL CARD

From the original Q-Signal usage, QSL was defined as "I acknowledge receipt" or in question form "Do you acknowledge receipt". In today's Ham usage, it is can be taken several ways. In a conversation, it is often taken to mean "I copied OK" (Did you copy OK?) or "I understand" (Do you understand?). Another meaning is to confirm a two-way contact between Amateur Radio Stations. In this sense, QSLing is exchanging QSL cards via the mail service or QSL bureaus. The cards are used to verify the requirements for a variety of awards including DXCC, WAZ, WAC, and many others. QSL cards were used early in the development of radio dating back to 1919. This chapter will address the means and methods to send and receive QSL cards.

To QSL or not to QSL is a question you might consider before getting too far into it as it can become expensive. You know that you successfully made a two way contact and that may be enough satisfaction. You may want to consider the new electronic QSL system, which is essentially a no expense way to get confirmations. However, these do not count for DXCC credit or other awards at the present. See Electronic QSLing later in this chapter.

9-3. LOGGING

You should keep all contact information in a notebook or logbook and a computer if you have one. There have been several incidents of fire, hard drive crashes and the like where years of data have been lost. It is a good idea to make zip discs and copies of paper logs and store them elsewhere. Essential logging information includes:

1. Date and Time (UTC)
2. Frequency or Band
3. Mode
4. Time Off (UTC)
5. Reports – sent and received
6. QSL Via Direct __ Manager ___ E-Mail ____
7. Contest Serial Numbers
8. Stations Worked - immediately before and after your contact

Optional Items are:

9. Power
10. QTH of Station Worked
11. DX Operator's Name

There are dozens of logging programs available for DXing, contesting and general logging – see URL's http://ac6v.com/logging.htm Many of these are free.

Needless to say, keep all these records indefinitely - your DXCC submissions may get lost - or worse a fire gets all your hard-earned QSL cards. You never know – what if years later you decide (heaven forbid) to work all counties. Also a DX station may request a QSL card years later when they find out you have a rare prefix or are in a rare county or grid square. Also you may want to goldmine your logs, even years later. Go back over your logs periodically and check for that rare one where you sent the card, but never received a return. I have had cases where a re-submission 3 years later yielded a card!

9-4. QSL CARD DESIGN

It is probably best to have two types of QSL cards, one for DXpeditions and rare DX countries and another for all other QSLs. It is doubtful that most DXpeditions will care about your 10/10 number, county, grid square etc., but many individual DX stations will want the details. Following are two examples of QSL cards for DXers. Avoid script or other unusual fonts that can be misread.

All QSL cards must contain the following information on one side of the card. QSL Managers and DX stations find it unnecessarily time consuming to turn cards over to find your call or QSO information. If you insist on a two-sided card, include your call on the same side with the QSO information:

1. Your callsign, name, address, country, zip code, and E-mail address.

2. Use an E-mail address that is independent of your provider (e.g., amsat.org, arrl.net, hotmail, yahoo mail, etc.), so that your card does not become obsolete should you decide to change providers.

3. Callsign of the Amateur Radio Station contacted, callsign of the QSL Manager if applicable.

4. Date and time of contact. This is the UTC date. Since various countries use different notations (i.e., 10/1/99 or 1/10/99 for October 1, 1999) it is best to spell out the month followed by the day. Many use roman numerals for the month e.g., March = III.

5. Band or frequency of contact. Mode of operation, SSB, CW, RTTY, etc.

6. Signal report exchange – in the RST system. See Appendix A3.

7. Contest name and contest exchange number if applicable.

8. Check boxes for "Please QSL" or "Thanks for your QSL".

9. Your signature.

10. For a general all purpose card you can add information used for awards, such as:

11. 10/10 Number for 10-meter contacts. Ten-Ten International was established to encourage 10 meter activity. For 10/10 number applications, see the Glossary.

12. County, province, prefecture, oblast, or parish.

13. Grid square – see Glossary. Many JA's collect these on HF

14. Zone – both CQ and ITU – see Appendix A6 and A7

15. Your awards, especially those that count for reciprocal award working

16. DX Clubs you belong to. Many offer awards for working a specified number of club members.

QSL VIA DA BURRO?   IRCs?   USE GREEN STAMPS

| GUD |
| Most places |
| One IRC |

A typical all-purpose card may not look like this but contain the following information

AC6V
CONFIRMSA TWO WAY QSO WITH
JA1BK

DAY	MONTH	YEAR	UTC	MHz	MODE	RST
25	March	2000	2057	28.480	SSB	59

DXCC SSB, CW   WAZ 10M SSB   WAS CW, QRP, 6-BAND

Rod Dinkins   Ten-Ten No. 18029   Kenwood TS-870
4982 Marin Drive   Grid Square DM13IF   Cushcraft R-7000
Oceanside, CA, U.S.A   CQ Zone 03, ITU 06   Ameritron AL-80B
San Diego County   San Diego DX Club   Kenwood TS-60
Zip Code: 92056-4973   San Diego PARC Club   QRP 6 Watts, QRO 1kW
E-Mail: ac6v@amsat.org   Website; www.ac6v.com
PSE QSL _X__   73 Rod Dinkins   TNX QSL____

A TYPICAL QSL CARD FOR DXPEDITIONS OR RARE DX MIGHT LOOK LIKE THIS:

AC6V
CONFIRMS A TWO WAY QSO WITH
TO0TX
QSL MANAGER OH2BN

DAY	MONTH	YEAR	UTC	MHz	MODE	RST
27	March	2000	2057	28.480	SSB	59
28	March	2000	1947	21.025	CW	599
28	March	2000	2108	24.945	SSB	59

AC6V E-Mail: ac6v@arrl.net
Rod Dinkins
4982 Marin Drive
Oceanside, CA 92056-4973, U.S.A
PSE QSL _X__ 73 Rod Dinkins   TNX QSL____

Note that many overseas QSL cards, use this format:

DAY	MONTH	YEAR	UTC	MHz	MODE	RST
25	III	01	2057	28.480	SSB	59

17. Another QSL date format is 10/11/01 which is 10 Nov, 2001 which can be confused with 11 Oct, 2001 depending on the part of the world you are in, so it is recommended to spell out or abbreviate the months as in the above examples. Many use roman numerals for the month e.g., March = III.

It is an absolute necessity to have an accurate 24-hour clock in your Ham Shack that gives accurate time to the second in Coordinated Universal Time (UTC). UTC is translated from the French *Universelle Tempes Coordinate'*, thus the "UTC" abbreviation. This was formerly referred to as GMT, also referred to as ZULU time by MARS and the military. By keeping your log in Universal Time, the time on your card will match the DX station's log and save countless hours of searching. In fact, some managers check a minute or two before and after and if not found - you get the infamous "NOT IN THE LOG" for all your trouble. No one wants to convert between local time and the time in another part of the world so we all use UTC. UTC time can be found at 10MHz as transmitted by WWV or WWVH.

Remember that the date changes at 00:00 UTC. This is early evening in the U.S. (4 p.m. PST and 7 p.m. EST). For example, when it is 4:01 PM PST, March 10 here in California – it is 12:01AM March 11 in England or UTC time.

18. As mentioned earlier, write out (or abbreviate) the NAME of the month. For example, if you work a station on March 1, 1999 and you put 3/1/99 on the card and then send it to an overseas country, there may be confusion as to whether the date is March 1 or January 3. Most countries except the United States abbreviate the date with the day, month, and year. Many use roman numerals for the month e.g., March = III.

Do not correct mistakes, but rather make a new card. If an altered card is submitted for award purposes, it might be rejected. On the QSL card and both envelopes, use ink that will not smear when exposed to moisture. Most ink-jet printers use ink that will smear when exposed to moisture.

Use the RST system for signal reports, 59 and S-9 are not the same thing despite common usage. Be sure your CW contacts include the tone i.e., 599. Cards for CW credit have been rejected because the tone was left out.

Submit all QSO's on one card in one envelope, either write them all in one area or use a label, or write a note telling where the QSO's are listed, such as that additional QSO's are on the back of card.

The card design can be as simple or as elaborate as you prefer, depending on your budget, from homemade to quality semi-glossy stock. Generic cards are available from several QSL card printers, several of which are on the Internet.

It is doubtful that DXpeditions or rare DX countries are particularly interested in an elaborate card. Many Hams have a generic or homemade card plus a showoff card, often from a DX Club, which can get a price break with large member quantities.

A review of the cards received by the ARRL Outgoing QSL Service indicates that most fall in the following range:

Height = 2-3/4 to 4-1/4 in. (70 to 110 mm) Width = 4-3/4 to 6-1/4 in. (120 to 160 mm)
The IARU Region 2 has suggested the following dimensions as optimum:
Height 3-1/2 inches [90 mm]   Width 5-1/2 inches [140 mm]

- **MAKE YOUR OWN QSL CARDS   http://ac6v.com/dealers.htm#CRE**

- QSL Creator - Create Your Own QSL Card -- Downloadable Program -- Click on DOWNLOADS -- From "The Loaded Dog"

- QSL Maker - From WB8RCR

- QSL Cards - Make ur own from WA7S

- QSLMaker 2.2 From rsars.org - UK
  QSL card.com - Create and distribute your own online QSL Cards via the web!

- QSL Card Kit - W7NN QSL Kit. Publish your own QSL Cards using your PC and ink jet printer. See QSL Kit.

  AC6V's InkJet Card. Select an image (I used PowerPoint), visit a paper supplier and choose heavy stock with the preferred color. Get samples and see if they will go thru your printer. Print image on one side. On the reverse side print the QSL data -- call, time, RST, etc. Print several images per sheet. Trim to size with a paper cutter. This is OK for a few cards, but will take you broke as inkjet cards take a lot of ink. For economical quantity, take an inkjet master, then Xerox several on a sheet, cut to size. Really best to Xerox them as inkjet will smear and run when wet. See sample http://ac6v.com/brag.htm#QSL

AC6V's Photo QSL Card. Use your 35mm Camera -- take your favorite photo -- Shack, Mug Shot, Local Attraction, etc. Have photo finisher run off 100 copies or so (try drugstore coupons). For callsign - use an Avery transparent label. For QSL data, use an Avery stick-on label, on the back of the photo. Print with ink-jet or laser. For AC6V's homemade card see Oceanside Harbor and the Star Of India at the bottom of the brag page, Sample at URL ---
http://ac6v.com/brag.htm#QSL

9-5. FINDING QSL ADDRESSES

Seldom will a busy DX station give their address during their QRZing. Rather, QSL by CBA (callbook address) or via my manager XXXX, or via the bureau (Burro) is common. Listen carefully to the DX station when you work them. Listen specifically for any instructions the DX has regarding QSL'ing.

It is not unusual for different managers to handle different modes or bands. Sometimes a guest will QSL direct or have his/her own manager.

Several means are available to determine QSL routes:

1. Radio Amateurs Callbook (Flying Horse)
2. Published Subscription Lists (The Go List)
3. Internet Callbooks
4. Internet QSL DataBases
5. Internet DX Newsletters
6. Internet News Groups
7. Internet DXpedition Logs
8. DX packet Cluster
9. QSL Request By E-mail
10. DX Nets
11. Internet Search Engines

If you are going to QSL a lot of DX by direct address, it is worth the expense of buying the Radio Amateurs Callbook (RAC), now available only on CD ROM for computers.

For DX Routes and QSL Managers, the GOLIST and several other published manager lists are available. See URL: http://ac6v.com/callbooks.htm

For those on the Internet, there are dozens of country call books (but by no means all countries). See URL: http://ac6v.com/callbooks.htm

On the Internet, you can find many QSL databases such as Pathfinder, Internet DX bulletins such as the 425 DX news, Ohio/Pennsylvania Bulletin and several others. In addition many DXpeditions now have online logs where you can verify your contact.

Check the DX Packet Cluster – use SH/QSL.

After you have conducted a search via these vehicles, one can always ask on the newsgroups or QSL reflectors. Try   rec.radio.amateur.dx news group. Several on-line QSL mail reflectors are available as well.

The International DX Association (INDEXA) provides QSL routes via the daily information net conducted on 14,236 MHz at 2330Z and their quarterly newsletter, see URL: http://www.indexa.org/ Surprisingly, using a web search engine such as goggle.com can also find QSL routes.

All of these Internet resources are free and can be found at URL: http://ac6v.com/ includes:

1x1 Calls	Electronic QSLing	QSL Galleries
10-10 Calls	GoList	QSL Managers
Antarctic QSL Page	Ham Populations	QSL Primer
ARRL In-Out Buros	INDEXA	QSL Printers
BuckMaster	IOTA QSLs	QSL Reflectors
Bureaus, ARRL In-Out	IRC Tips & Advice	QSL Services
Bureaus, World Wide	IRCs Required - To USA	QSL Tips & Advice
CallBooks - World	LogBooks - OnLine	USA Call Books
CallBooks - USA	Old Call Lookup	Vanity Calls
Contest Logs - OnLine	Postal Rates	Vanity Call Applications
DX Callbooks	QRZ.Com	WM7D's QSL Lookup
DX Managers	QSL Buros - World	YASME QSLs
DXpedition Logs	QSL Queries By E-Mail	

9-6. QSLING SWL STATIONS

It is important to respond to SWL (shortwave listener) cards; in some countries it is the entry into Amateur Radio. To respond, fill out the card with the SWL callsign in the slot normally used for the station worked, include the date/time/band/mode info the same as if for a QSO except for the mode entry, write in "2xCW with EL9XX". Send the cards via bureau and mark them SWL RADIO XXXX (e.g., SWL-QQ12345). Several QSL printers offer generic SWL cards. See URL: http://ac6v.com/dealers.htm#QSL

NO NO Guglielmo –
I said lower the tilt-over tower!!

9-7.   SENDING QSL CARDS

There are several ways to send QSL cards:

1. Direct to the DX station by domestic or international mail. Paragraph 9-8.
2. Via a QSL Manager. Paragraph 9-9.
3. Via your QSL Bureau. Paragraph 9-10.
4. Direct to the DX Bureau of the country worked. Paragraph 9-12.
5. By using a QSL Service. Paragraph 9-13.
6. By E-Mail (some managers will accept your E-Mail and if in the log – will send a card via the bureau), not too common but is being done.
7. Electronic QSLing. Paragraph 9-14.

If the QSL route is via a manager, you have two options – QSL via the manager's address or dump a card into the bureau and hope for the best. Not all QSL managers will return a QSL if it arrived via the bureau. If the QSL Manager is in your country, it's a matter of simple domestic mail. For example a lot of USA operators contest from the Caribbean, but QSL from a stateside address.

Beginning DXers are oft tempted to QSL direct for the common countries (e.g., most of Europe, Japan, Central and South America). To save money over the long run (332 direct QSLS could cost you over $600), use the via-the-bureau method – eventually you will confirm them but it takes time. Just work several stations from a given country, one or more will eventually come through. If in doubt ask a DX Elmer what is common and what is rare. See Appendices A12 and A13. Note that some countries do not have DX bureaus.

More than one DXer will advise that if you need a particular card, sending direct is best way using nested printed envelopes with foreign return postage enclosed. Expensive but gets results. QSL services like WF5E is a close second and a lot cheaper and simplifies things considerably. The bureau is fine for run of the mill cards, those considered to be common DX.

9-8.   SENDING QSLs DIRECT

The fastest response is usually by sending an SAE and postage directly to the DX station. The preferred method of addressing in order is: typed addresses, computer-printed address labels, and **hand printing in all capital letters**. To minimize looking like QSL mail, use a pre-printed business like envelope and make your return address appear business-like. Do not use script writing. Try to avoid thickness in the envelope. Perhaps use the XYL's name as well -- Rod and Karla Dinkins may appear more a personal envelope rather than a QSL type.

If you have an acquaintance that knows about the country you are QSLing, have them put their cultural touch on the addressing. Marking the envelope "No Valuables Inside" in the native language may help or defeat your purposes.

Maybe use your degrees as part of the return address MD, MS, AA, may command respect and of course never include callsigns – yours or the DX callsign. Mail theft is a fact of life in some countries, so it is best to not use your callsign on any overseas correspondence.

On the other hand, envelopes that are well sealed, opaque, or thick might attract more attention than a routine-looking method. Who knows? You might have to try both methods. More than one DXer has sent QSL cards over and over again before success. Addressing for some countries might be confusing. For example, from the RAC callbook:

Pete Harris, G0WUA
1 Newlands
Landkey
Barnstaple
Devon EX32 0NJ
Great Britain

This can be placed on a standard computer label (max 4 lines). Just be sure to put in the commas as they appear from the RAC address.
Pete Harris, G0WUA
1 Newlands, Landkey
Barnstaple, Devon EX32 0NJ
Great Britain

9-9.   QSLing VIA A DX MANAGER

If a DX station states QSL via my manager they mean instead of sending it directly to the callbook address (CBA). If you are in the same country as the DX Manager, it's a simple matter of preparing your card along with a return self-addressed & stamped envelope (SASE) and inserting these in a larger envelope and mailing it to the manager.

Usually you will include all QSO's (bands and modes) made with the DX station on one card (instead of sending 3 cards if you work the DX station on three different bands or modes).

However be aware that some managers require that you include a finite number of QSO's. An example is the recent D68C operation where the QSL Manager could only handle 3 QSO's per card due to his QSLing program. Check the DXpedition website or the DX newsletters.So best check on the web for requirements, most big DXpeditions will have a web page set up. Some even have a "log check" where you can check to see if you are indeed in the log. Use google.com to search for a DXpedition web page or use the QSL resources at ac6v.com/callbooks.htm.

If the QSL is urgent, request only one QSO with one DX call. If you request multiple QSO's, you run the risk of not all QSO's being in the logs (not all logs may be with the manager). This requires further research and the manager may set it aside until the necessary logs arrive.

When the manager receives your card, it goes into a stack awaiting processing. How long it will take for processing depends on several factors. Processing cannot begin until the manager receives the logs from the DX station. Logs might be emailed, sent by airmail on paper or disc or delivered by some intermediary, or transmitted over HF voice or packet. The manager may be waiting for QSL cards to be printed.

If you have worked a DX station that has a foreign manager, you have several choices:

1. Send your QSL card and envelope directly to the manager with postage remuneration.

2. Send your QSL card via a QSL Card Service. If you're not in a hurry, you can use a QSL card service to forward your card to the manager.

3. Send your card via the bureau. Write on the card the DX station's call via the manager's call.

For Stateside QSL Managers, on back of the envelope, print your call, date and time of contact, and the band. Many organize their cards by batches to facilitate handling. For some overseas DX Managers, you may want to do the same, but be aware it is a tip off to postal thieves.

A well known DXer and his QSL Manager advises as follows: No insurance contacts please, check the on-line log book, if you appear in the log on a specific band and mode, don't call us again. This will help us to work more individual stations, especially those running low power and antennas. It may take up to 48 hours for the log to be updated after your contact. Only if you are really (really!) uncertain we logged you correctly, please wait and check the log for a couple of days before calling again. This to avoid an un-necessary duplicate contact, which will have the effect of blocking a contact with someone who may still be trying for their first ever QSO.

Some QSL Managers handle QSL chores for several DX operations and prefer that cards for different operations be sent in separate envelopes. The reason being that sometimes they run out of cards for one station and have to wait for a new supply from the printer or are waiting for arrival of the logs.

Be patience. Cards do not return overnight. Expect a minimum of 10 days for domestic cards and 4 weeks for overseas cards. In the cases of a major DXpedition, expect months for a return, since most of the time (and there are exceptions) cards are not printed until after the Dxpedition returns and the managers literally have tens of thousands of QSL requests to check through.

9-10. OUTGOING QSL BUREAUS

The outgoing QSL Bureaus are clearinghouses for QSL cards offering a convenient and inexpensive method of QSLing. Many countries offer these QSL Bureau services. One of the greatest bargains of League membership is being able to use the ARRL Outgoing QSL Service to conveniently send your DX QSL cards overseas to foreign QSL Bureaus. Your ticket for using this service is proof of ARRL Membership and just $4.00 per ½ pound. (Those not quite so DX active can send 10 cards [or fewer] and enclose $1.00 And the potential savings over the substantial cost of individual QSLing is equal to many times the price of your annual dues. The Service handles approximately 2,000,000 cards each year! The ARRL Outgoing QSL Service serves approximately 260 DXCC countries. Even Hawaii and some US possessions. For countries served, see URL: http://www.arrl.org/qsl/qslout.html

No envelopes or postage remuneration are required or allowed other than the above fee. For example, if you work 150 DX stations, and you send direct for each one, this might cost upwards of $2.00 to $3.00 each for your postage and included return postage or about $300.00. Sending via the bureau would only be about $8.00 or so total for all! A rough rule of thumb is that approximately 75 cards weighs ½ pound.

QSL cards are shipped to QSL bureaus throughout the world, which are typically maintained by a national amateur radio society for the particular country. These incoming bureaus sort the cards and volunteers stuff envelopes for further distribution to the individual Amateurs. The ARRL QSL Service cannot be used to exchange QSL cards within the 48 contiguous states. For complete details of the ARRL Outgoing Bureau and the latest costs – write the ARRL or see URL on the Internet: http://www.arrl.org/qsl/qslout.html

The big disadvantage of the bureau system is that it can be slow, taking up to a few years to send your card and get the DX station's card back. If you are anxious, and want the card quickly, you can QSL direct as covered later in this chapter. However, QSLing direct can be expensive if you intend to do a lot of DXing, the bureau systems are very cost effective.

You can send a card to a QSL Manager as well, just label it as (DX Station) via (DX Manager), e.g., D68C via G3SWH and sort it in the G group.

Beginning DXers are tempted to send direct to countries with fair size ham populations such as Japan, Brazil, Argentina, etc. However this is expensive and you will be better off to work a dozen or so from each country and use the ARRL Outgoing Bureau, you will receive replies eventually.

Note that some countries do not have QSL Bureaus, so check your card file against the list at Appendix A-12.

As the cost of burro postage increases, some managers or DX stations refuse to process cards from the bureau, or hold them off for months. It is difficult to blame the managers or the DXers who can't afford to go broke handling the costs involved in sending or receiving burro cards. Some of the well-known QSL managers will attest to this. So if it is an all time new one, a direct mailing to a manager or DX station with sufficient return postage is in order.

9-11.   INCOMING QSL BUREAUS

Now that our QSL is on its way via the slow burro, how do we get a return?? Most countries have "outgoing" QSL bureaus that operate in much the same manner as the ARRL Outgoing QSL Service. The members send their cards to their outgoing bureau where they are packaged and shipped to the appropriate countries.

Within the US, the ARRL Incoming DX QSL Bureau System is made up of numerous call-area bureaus that function as central clearinghouses for QSL cards arriving from foreign countries. Volunteers staff these "incoming" bureaus. Except for postage costs, the service is free; ARRL membership is not required. A majority of the DX QSLs is shipped directly to the individual incoming bureaus where volunteers sort the incoming QSLs by the first letter of the call sign-suffix. One individual may be assigned the responsibility of handling one or more letters of the alphabet. Some incoming bureaus now have web pages – see ARRL pages.

Note that some of the bureaus are very fussy about the size of the envelopes that are kept on file. Many will accept a check and set up the envelopes for you. Your check will cover envelope and postage costs. To be sure what your district bureau wants – write to them first. Generally, you should supply them with address labels along with a check. For more information on using the ARRL Incoming QSL Bureau, write the ARRL or see URL: http://www.arrl.org/qsl/qslin.html

9-12.   QSLING TO A DX BUREAU DIRECT

Many countries have Incoming QSL Bureaus should you prefer to shorten the time for a QSL return. No self-addressed-envelope (SAE) or self-addressed-stamped-envelope (SASE) is required, as they will return your QSL via your Incoming bureau. For a list of the IARU QSL Bureaus – see URL: http://www.iaru.org/iaruqsl.html   If you have a lot of cards to send to a particular country, check the various rates, small package, etc. It might be cheaper and quicker to send cards direct to a DX bureau than sending via the ARRL outgoing bureau. Mark the package "QSL Confirmation Slips" to avoid the impression of sending post cards.

9-13.   QSL FORWARDING SERVICES

The WF5E DX QSL service is a well-known forwarding service to obtain QSLs. It saves the hassle of keeping multiple QSL information, finding out who to QSL to, and sending out your own SASE's. The QSL service has been in operation almost 37 years, serving approximately 5000 customers in every state and over 50 countries. They receive 8000 to 9000 QSLs per month to forward. Service includes forwarding QSLs to DX and USA managers, active DX stations, and DXpeditions, which consist of over 4000 stations. No stateside to stateside QSLs are handled. For more information see URL: http://www.qsl.net/wf5e/

9-14. QSLING ELECTRONICALLY

A new innovation is occurring on the internet – electronic QSLing, the Electronic QSL Centre at URL: http://www.eqsl.cc/qslcard/ has processed over 3 million QSL "cards" and the service is free although donations are in order if you prefer. **Be aware that DXCC and other awards are not currently accepting these for credits**, but may be in the future. So if you are not after a mailed QSL and the awards with their attendant costs, electronic QSLing may be for you. Getting QSL cards direct or via an overseas manager might cost $200 - $300 for DXCC or $1000 if you get to honor roll. But nothing beats the beauty of a QSL card from distant lands and they make a nice brag item to civilians. Also check the new ARRL Log Book Of The World – URL: http://www.arrl.org/lotw/ this program does count for award credits.

Some awards will specify **GCR** which means General Certification Rule - Most sponsors allow GCR in lieu of actually wanting to see your cards. You need to have the cards! GCR usually means getting the signatures of two witnesses who certify that you possess the cards and that the information you state on the application is correct.

9-15. ENVELOPES

Mail theft can be a problem in some areas of the world. It is a real challenge to directly mail QSLs to hams in these areas as mail is routinely pilfered for valuables such as US dollars. There are 3 basic ways of mailing postage overseas:

1. IRCs - International Reply Coupons
2. Green stamp – USA one dollar bill, Euro Dollars are now also being used.
3. SASEs - self-addressed stamped envelopes with proper country stamps.

9-16. ENVELOPE SIZES

The US #6 personal size (3 5/8 x 6 1/2 inches) is considered a poor choice for overseas mail. The US #10 size (4 1/8 x 9 1/2 inches) is better and it will accommodate the #9 return envelope without folding. However the best overall choice is the European mailer (4 3/4 x 6 1/2 inches/ 120 x 165 mm.) This will accommodate the European return size envelope (4 1/2 x 6 1/4 inches/ 115 x 160 mm).

Most managers recommend the opaque or "security" airmail striped envelope. In addition, using an envelope with a pre-printed business return address may help disguise the much sought after QSL package. DX Managers constantly recommend using envelopes that can easily hold the standard 9x14 cm QSL card, i.e., use envelopes which are at least 10x15 cm. Dry glue envelopes (peel and stick type) are recommended for damp weather environments.

Use the ones with a strip that is removed before the envelope is sealed. Bill Plum sells a very nice set of European Airmail Envelopes. One fits inside the other with no folding at all. According to his last catalog, the outers are $9/100 and the returns are $8/100 + shipping. His fax # is (908) 782-2612.

Send an SAE or SASE not just an address label. **Put the folded end of the return envelope on the bottom of the mailing envelope.** This will prevent it from being sliced when opened by a machine or letter opener. On your return SAE or SASE, make sure your address is readable and complete including your country and allow sufficient space for a postage stamp.
Write the managers address in the return address area of your return envelope, this saves time and allows for a return path should that be necessary.

Insert your QSL card in the envelope with the card facing the flap. **Do not insert a Green Stamp or IRC inside the return envelope as this causes extra work for the manager.** Stuff in this order: SASE or SAE, Green Stamp or IRC, QSL card. Do not seal the outer flap of the envelope all the way to the very edge, thereby allowing a small space to easily open the envelope.

Test your entire package with a 150-Watt flood lamp. If Green Stamps or IRCs are evident, use a piece of carbon paper, cut to the size of the larger envelope, inside at the front of the envelope facing the front. All the other contents are placed behind the carbon paper to avoid smearing. You can also use filler sheets cut up from scrap paper to disguise the envelope contents. Better still use one of the many varieties of "security envelopes" available at office supply stores.

There are two basic strategies for getting your envelope safely delivered: concealment or open.
Concealment seems to be the most commonly used where there are no Amateur Radio calls on the envelope and an effort is made to disguise the contents. Many use an opaque or security envelope, keeping the package as flat as possible. A printed "business-looking" return address in the upper left corner may help.

Return postage can be IRCs or **new** $1 or $2 bills not folded or wrinkled. Also you can buy foreign stamps and send an SAE with the proper return postage. Don't affix the stamps – managers prefer them enclosed but not glued.

Open method. Use the European-size mailer and a slightly smaller return envelope, privacy-lined, red/blue airmail border envelope flap tucked in, **not sealed!** Inside, place the return envelope with the foreign postage. A high rate of return has been claimed for this method. Hopefully, if mail thieves peek in the envelope and see the SASE, they will allow it to continue on its way. Thieves that are attracted to a sealed envelope will save it and later destroy everything to avoid leaving any evidence.

IRCs although fairly universal cannot be redeemed for postage in some countries. They are sometimes traded as "ham currency", however SASEs are much easier for the DX operator.
Sending QSLs via registered mail may be viable - sometimes works, sometimes not. Some DX operators may specify registered mail in the QSO.

Another possible stealth method is cutting off the lower, left-hand corner of the mailer envelope about 1/2 inch diagonally. This creates the impression of mail sent at an economy "printed matter" rate. Also marking the envelope "Printed Matter" with a rubber stamp may help.

A QSL Manager advises: PLEASE DO NOT use #10 business envelopes for QSLs. If you insist on a #10, use a standard envelope for your return card, and fold it in half pointed downward, so it doesn't get cut in half, or have the flap cut off. With a letter opener, the triple folded #10 ends up in 2 pieces!!

This requires excessive time to tape back together!! Nesting envelopes are sold at many stationary stores and allow you to stuff the QSL package rather than make multiple folds on the return envelope.

When addressing, do not use your call letters or those of the DX station on the envelope. This is a signal to postal thieves that money may be enclosed. Also do not use attractive stamps on the envelope; they may be desirable objects in certain countries. Perhaps have the post office imprint the postage, thus avoiding stamps altogether. Strive to make your package look and feel like a simple pen-pal letter you are writing to a friend in a foreign country. Or make it appear as a business-type letter with biz like return address. Using nesting envelopes will reduce the bulk and appear as a normal letter.

9-17.   RETURN POSTAGE

Return postage can be made in several ways:

1. Cash (The Green Stamp)
2. IRC
3. Stamps
4. Checks

A problem arises when certain countries have different airmail rates for various parts of the world. For mail to some countries, this rate exceeds the amount for the minimum airmail rate. A QSLing approach that has worked well is to use one IRC for Europe (except Germany) and a Green Stamp (s) for other areas that do not prohibit US dollars.

9-18.   POSTAGE STAMPS

In most cases, including postage stamps for the country that you are QSLing is a sure bet. Be sure to send sufficient face value in stamps to ensure return. Several vendors on the Internet sell foreign stamps for virtually any country in the world. Managers recommend that you **do not** affix stamps to your SAE, but rather include them in your envelope. If you have sent too much postage, the manager can use the extra value for printing costs etc.

9-19.   CASH

Except for the US dollar bill (Green Stamp), most managers have no use for foreign currency, requiring a trip to the bank and extra fees for exchange, if indeed it can be exchanged. However Euro Dollars are starting to be used. The US Green Stamp is fairly well accepted around the world, however be aware that in some countries, Green Stamps are prohibited – in fact the DX station may get into trouble if you send them.

Although the author has never seen such a list, it is fairly easy to figure out by virtue of the political climate between the US and other countries, which countries to avoid sending Green Stamps.

Banks in some parts of the world charge large fees to convert US or other currency.

So the DX may have to wait until they have enough on hand to make the conversion and related costs worthwhile. Green stamps are fairly easy to obtain by DXers in western countries, they are useful in many countries, less expensive than an IRC, and they save trips to the post office. When sending green stamps, send new ones. Some foreign administrations or banks can be funny about handling older bills. When including more than one US dollar for return postage, rather than send two $1 bills, send a $2 bill, which will cut down on the thickness and weight.

When the US dollar is strong, the DX station may get more than its value compared to an IRC. On the other hand, the exchange cost may make a single Green Stamp insufficient for return postage. If in doubt, ask the DX station perhaps by E-Mail, or ask another DXer to determine the number of Green Stamps required.

In countries with unstable currencies, the correct amount of Green Stamps may vary. For example in Germany, where the DM rose dramatically against the dollar, it requires more than one Green Stamp to buy postage. At present this doesn't seem to be the case with much of the rest of the world. With the advent of many DXers and managers now on the Internet, it is quick and easy to E-Mail the DX or manager.

9-20. BANK CHECKS

This method is rarely used because of the complication of cashing it in a different country. However there have been some instances where the submitted check had QSL information on the back – thus verifying the QSO when cashed!!

9-21. IRCs

Often abbreviated to IRC in English, C.R.I. in French (Coupon-Reacuteponse International), and IAS (Internationaler Antwortschein) in German, the International Reply Coupon was created by the Universal Postal Union. These allow a correspondent in any one country to pay for the postage required to cover a reply postage from the receiver of their letter in another country. An advantage is avoiding any exorbitant bank charges or possible exchange control regulations.

The coupons are redeemable for postage stamps in any country, which is a Member of the Universal Postal Union. Here is a list of countries who do not accept IRCs: Abu Dhabi, Anguilla, Antigua, Netherlands Antilles, Bahamas, Bahrain, Bhutan, Bolivia, Brunei, Cayman, China, C.I.S., Czech Rep., Dominican Rep., Dominica, Falkland, Gilbert, Grenada, Guyana, Honduras, Korea North, Korea South, Lebanon, Lesotho, Nepal, New Hebrides, Montserrat, Oman, Peru, Qatar, Saudi Arabia, St.Kitts & Nevis, Saint Lucia, Saint Vincent, Swaziland, Taiwan, Tonga, Hungary, Vatican City, Vietnam, Virgin Island, Yemen, Zimbabwe. However, IRCs are sometimes accepted by hams in these countries, because they can use them as return postage for their own letters.

Despite the international nature of IRCs, many post offices around the world frequently fail to keep IRCs in stock. Also the IRC invariably costs the purchaser somewhat more than it delivers to the recipient in terms of postage. The balance of the monetary value involved is simply taken by the post office as a charge or commission for the service provided.

The space on the bottom left of an IRC is for your local post office to date the coupon to validate it (the spot where the purple postmark is). The center space is the local cost and the right space is for the recipient's post office to place a validation mark.

In the United States, each IRC is currently $1.75 as of January 7, 2001. See USPS Publication 51 at URL: http://www.usps.com A one-ounce letter sent airmail to any world destination (except Canada and Mexico) will cost 80 cents. A two-ounce airmail letter will cost from $1.55 to $1.70, depending on the destination. The complete rate schedule is available at the USPS Web site. Redemption can be a variable since post offices in some countries don't respect the international agreement, often redeeming an IRC for about 80% of the face value, keeping 20% for processing fees. For example, in Tahiti, for 1 US IRC, redemption is only 80 cents in stamps, not sufficient for one airmailing.

Probably the real value of an IRC for the QSL Manager is the price he can get by selling them to other Hams. As long as the going rate reaches or exceeds the current maximum postage cost in any country, the IRC is a viable approach.

An example of QSL-card costs sending a card to a German DX Manager:

1. $.05 for each envelope
2. $.80 for postage to Germany
3. $.05 for the card
4. $2.00 for the German stamp to get the card back
5. $2.90 total

If you buy used IRCs, be aware that depending on the date shown on the IRC, the exchange rate may be less than 60 cents and in some cases as low as 40-45 cents.

Not all (or even many) postal clerks are familiar with IRCs. Some refuse to bother with them. Some just give you an airmail stamp for each one no matter what it looks like, some examine every little detail and throw out any they can't read or are not appropriately marked.

9-22. MONETARY TIPS FOR QSL MANAGERS

Well one thing for sure: the QSL Managers put in a lot of voluntary time opening, reading and returning QSL cards. Printer costs for QSL cards are high and many hours are involved handling thousands of QSL cards. I doubt we could achieve Honor Roll without them. A small gratuity is certainly in order to help defray costs and show appreciation for their efforts. Some managers may appreciate pristine collector stamps as well.

Advice from a QSL Manager -- "It would also amaze you the number of people who send a card direct but do not include an SAE or any return postage whatsoever, yet seem to think that I am charitable enough to return their card direct - sorry - they go back via the bureau. If you are not a member of the buro, that's your problem!
I am not a charity nor do I expect or wish to make any money out of QSLing. I do it solely because I too like to receive cards and know sometimes it is difficult to get cards".

9-23. CURRENT US POSTAL INTERNATIONAL RATES

Since the postal rates change quite often, best check the US Postal schedule at URL: http://www.usps.com/welcome.htm

9-24. WHAT IS A GOOD IRC?

For an IRC to be valid, it MUST NOT be stamped in the right hand box. IRCs have become a de facto Ham Radio currency, and there are many improperly stamped IRCs in circulation. If you buy new ones, be sure your postal service stamps them correctly. If you buy from another Ham, check that they are properly stamped.

AIRMAIL IRC. Check that you have an airmail IRC. You can determine this by looking at the front of the IRC in the paragraph above the three boxes; the last three words should say, "par voie aerienne". Airmail!

LEFT BOX. The IRC must be stamped by your post office ONLY in the left box. This box has "Empreinte de controle du pays d'origine (facultative)". The "cancellation" appearing in the left box must have the date of issue, and be legible.

CENTER BOX. The center box shows the price paid for the IRC. Your post office may stamp the center box, but only to indicate the price of the IRC, not the 'postmark' stamp used in the left box. In some instances, the amount is already printed. If the center box is blank this has not been a problem in most cases.

RIGHT BOX. DO NOT allow your post office to place any mark in the right box. This cancels the IRC, indicating it was exchanged for postage. Beware of IRCs with a postmark in the right box; a post office may not redeem them.

9-25. HOW MUCH POSTAGE TO SEND

Here is the consensus of several sources. Not guaranteed but for what it's worth. Also see URL: http://www.k4hb.com/
All countries not listed are in general --one IRC or one Green Stamp
Argentina - $1.25 to mail to any American country, and $1.50 to send to any other continent
Australia   one IRC or two Green Stamps (costs have gone up so better ask)
Belgium (air) two IRCs or two Green Stamps
Easter is   one IRC or two Green Stamps

Germany   two Green Stamps, some sources say 3 IRCs
Lebanon   IRCs not valid, one Green Stamp
So. Africa   two IRCs or one Green Stamp
India   no Green Stamps
Switzerland   two IRCs or two Green Stamps
Uruguay one IRC or two Green Stamps
Direct QSLing into Russia -- all letters sealed with transparent tape are supposed now to be "suspicious", especially stateside ones. Most of them come opened. Dollar bills do not come through, IRCs are OK. Dennis, RZ1AK

9-26. QSL CHECKLIST

____ QSL Card has your E-Mail Address
____ If sent internationally, don't forget to include "USA" on the return envelope.
____ Contact Date in the form March, 17, 2000 – Or March = III.
____ Your Callsign and QSO information on the same side of the card
____ Indelible Ink
____ All QSO's on one card in one envelope, don't send a card for each band and contact.
____ Envelope Addressing use typed addresses OR
____ Computer-printed address labels OR
____ Hand printing in all capital letters
____ Pre-printed business envelope and a business-like return address
____ Envelope stamped "No Valuables Inside" or "Printed Matter"
____ Cut Corner Envelope
____ Opaque Security Airmail Envelope
____ European mailer (4 3/4 x 6 1/2 inches/ 120 x 165 mm.)
____ European return size envelope (4 1/2 x 6 1/4 inches/ 115 x 160 mm)
____ Dry glue envelopes (peel and stick type)
____ Folded end of the return envelope on the bottom of the mailing envelope.
____ SAE or SASE not folded
____ On mailing envelope – flap has small-unglued space for letter opener
____ On return envelope – managers or DX station return address
____ Affix a USPS Label 19B sticker to the SAE
____ Stuffed in this order: SASE or SAE, Green Stamp or IRC, QSL card.
____ QSL card in the envelope with the card facing the flap
____ Test package with a 150-Watt flood lamp
____ Carbon paper, inside at the front of the envelope facing the front.
____ Return postage stamps attached but not glued to return envelope
____ IRCs stamped and valid
____ Green Stamp or Euro Dollar is new, unwrinkled and not folded
____ $2.00 bills when needed - instead of two $1.00
____ Tip for QSL Manager
____ For Overseas QSL Managers, see above, but be aware it is a tip off to postal thieves.
____ No callsigns on the outside of the envelope outside of USA
____ For Stateside QSL Managers, on back of the envelope, print your call, date and time of contact, and the band.
____ Print Asia, Europe, Africa, South Atlantic Ocean, South Pacific Ocean, Caribbean, etc. as the last line in the address on some of the more obscure islands/countries.

Note that Reunion Isle should be sent VIA FRANCE. Pitcairn Island should be addressed Pitcairn Island, South Pacific via New Zealand

3-27.   The Missing QSL Card

Not all QSL cards will come back in a reasonable period of time. Reasons could be your envelope was ripped off by postal thieves or got lost in the mail, or the postal clerk couldn't read your handwriting – use a printed address label. You included insufficient postage. Some managers take the insufficient postage QSLs and eventually dump them into the bureau system – could take a coupla years to get to you. Another reason for the missing QSL card is that somehow the logs got lost and all these contacts are unverifiable.

The manager or DX station is way behind – managers are essentially volunteers. The manager is waiting for a build up of cards before tackling the job, or waiting to print a new batch of QSL cards. And in a rare case, a poor manager who got in over their head, could take many months for them to catch up. In one case, the QSL manager died and it was a year before some one else obtained the logs. Could be a broken contact and you are not in the log – but most managers or the DX will inform you of this – but some don't. Perhaps you sent a card via the bureau to a manager and expect a reply via the bureau -- not all managers will honor this.

Some (but overall very few) DX stations just want the return postage money and don't reply, these get to be known and a post to the rec.radio.amateur.dx news group will usually get an answer if certain stations are indeed QSLing and who received theirs and when.

DX stations on temporary assignment may wait until they return home before QSLing, could be many months. Some DX stations have good intentions – but alas just are very poor at QSLing. In the last case, work another in the country and hope for the best. I have had many rare DX contacts and never received a return and some QSLs came in six years later !!!

Well what do? After about six months I send another card and adequate postage and make sure the envelope doesn't exhibit Ham radio content. I did this three times for a 3V Tunisia card via an Italian QSL manager – but I finally got it. Write a postcard or look up the manager or DX stations E-Mail on the web and ask – have them respond to your e-mail address -- most find this preferable to a letter. This has worked several times where I was informed that they never received my card. So I sent another.

Go back periodically and look at your logs, try again for those over six or twelve months old.

I said turn on the RADAR – to heck with the doves

**CQ CQ CQ   PA9DXX CALLING CQ ON 160 METERS
BEAMING 4th STREET, MIAMI, FLORIDA USA AND BYE**

APPENDIX A1 CW OPERATING PROCEDURES

APPENDIX CONTENTS

This appendix contains information on CW Operating Procedures. The table below lists the chapter topics and paragraph numbers where they can be found. Info on learning the code and passing the code test can be found at URL: http://ac6v.com/morseaids.htm#Learn

PARAGRAPH & TOPIC	PARAGRAPH & TOPIC
A1-2. CW Tutorial	A1-5. Iambic Keying
A1-3. Prosigns For Morse Code	A1-6. Adjusting Paddles
A1-4. CW Bandwidth And Character Spacing	A2-A7. CW Abbrev, Q-Signals, And RST

A1-2 CW TUTORIAL

In addition to the quickie DXpedition and contest CW contacts, you can work a lot of DX in a one-on-one QSO. A lot of common DX is available on CW, with very little competition. For example, the mob will pileup on a European station on phone, but on CW on many occasions you can hear that same country calling CQ with no takers. An excellent CW Tutorial can be found by Jack Wagoner WB8FSV at URL: http://www.netwalk.com/~fsv/CWguide.htm

Lets take it step by step.

1. Have a list of CW Abbreviations, Prefixes, and Q-Signals handy. Some DX stations cannot converse in English but you both can get the essentials across with Q-Signals.

2. Know how to "zero beat" a CW signal. Many stations have very narrow filters and you want to be in their bandpass. Refer to the operating manual for your radio. However one CW pro advises, "The assumption that replying at zero beat is highly desirable is incorrect. Good and competitive CW operators know that calling as much as 200 Hz off frequency is usually much more productive in pileup situations. The XIT control is useful for this. A disadvantage of razor-thin filter selectivity is that this very useful strategy does not work well. A good operator uses his "ear" as an adaptive filter and lets the receiver cover a few hundred HZ."

3. Know how to use your RIT, XIT, Dual VFO's, and CW filters. See Chapter 2.

4. Listen for a DX station calling CQ or wait until they have finished with a QSO.

5. Good operators will send KN as a turn over, which is "go ahead, over, others keep out." Sending just K opens it for others to break in and this is OK if that is desired. SK is the signoff that should be used or CL ("clear") if closing your station.

6. Give a call in 1 X 2 call format -- DX11DX   DE   WA6WTO WA6WTO AR (The AR is a prosign sent as one character, i.e. didahdidahdit and means that I am through with this transmission). The DX station knows their call, so send it once. Sending your call twice allows the other station to hear it, then confirm it.

7. If you make the connection, the usual follow up is his or her signal report, repeated twice if the contact is shaky and weak, then your name and QTH. Don't send more than that on the first round. Turn it back to the DX with a K or KN. This will allow the two of you to evaluate if a QSO is sustainable or desirable.

8. On the next over, ask about QSL information if you want it, before the band slips out.

9. If no DX is calling CQ, but the band seems open, find a clear frequency and listen for a bit, if clear, then send QRL? QRL asks is the frequency busy? If someone responds with C, or QRL, no need to respond and clutter up the frequency. If no response to your QRL, repeat a couple of times and then call CQ. Sending just QRL without your callsign is against the rules, but most do it anyway.

10. Calling CQ is typically in a 3 X 2 format   CQ CQ CQ   DE   WA6WTO WA6WTO K.   Long long CQ's are likely to be ignored. Don't use AR instead of K as it means ending the transmission, and not an invitation for an answer. KN is a turn over to the station you are already working in a QSO

11. Listen for a few seconds using RIT to check for off frequency responses. If you have a narrow CW filter in line, use RIT and tune up and down from your transmit frequency to determine if someone is responding.

12. Repeat your CQ or QSY to a clear frequency, as you may be on a Big Gun frequency that can't hear you.

13. After the initial contact, it is typically DX11DX   DE   WA6WTO GM (GA, GE) OM TNX   FER   CALL   UR   RST ###   (339, 599, 549, etc.) NAME HR IS ROD ROD. QTH IS SAN DIEGO, CA. SAN DIEGO, CA. HW? AR DX11DX   DE   WA6 WTO K (OR) KN. See your list of abbreviations if you are not familiar with these.

14. DX11DX returns with essentially the same info, you may get "R" indicating that DX11DX copied all, or QSL on all is sometimes sent.

15. The next round is an invitation to rag chew. If DX11DX is too fast for you, send a QRS (send slower please). Longer QSO's usually include your station configuration, the weather (WX), jobs, ages, etc. You may receive an invitation to operate QSK (break-in) where the QSO is much more conversational. Practice with a friend first as this takes some getting used to and proper equipment settings.

16. It is not necessary to do a (DX11DX   DE   WA6WTO) every time except every 10 minutes of course. When you turn it over – you can use BK or just KN or K.

17. A signoff looks like this:

DX11DX DE WA6WTO, FB VLAD TNX NICE QSO HPE CUL VY 73 GM SK
DX11DX   DE   WA6WTO Use SK or CL (Closing Station) on your final transmission
not AR or K (N)

Then there are some cuties signoffs   dit dit, and a response of dit. Old Military types
use dit dita dit dit (Shave and a Haircut) with the response of dit dit (2 Bits)!

18. For contests, a common CQ is "test AC6V test".

19. Tail-ending. Wait until another QSO is complete, and then call.

20. Breaking into a QSO is not commonly done on CW and should be approached with
caution. If it obvious that two old friends are in conversation, it is not advisable. If the
exchanges include KN – it's a signal that others are not welcome, best wait until the QSO
is over and then tail-end. The standard break-in method on CW is to wait between
transmissions and then send "BK" for break, or "BK de WA6WTO".

It is very common to send RST reports in abbreviated form, for example 599, is sent as 5NN.
"N" in place of the number "9". Also another time saver is for the zero using a long "T". "T"
is sent in place of the number zero as in " POWER HR IS 3TT WATTS". There is a number
code for all numbers; however, the N and T codes are the most common ones.
Also CW stations sometimes report their zones as "A4" or "A5" instead of sending "14" or
"15". 1 = A,   2 = U,   3 = V,   4 = 4,   5 = E,   6 = 6,   7 = B,   8 = D,   9 = N,   0 = T

A1-3   PROSIGNS FOR MORSE CODE

Prosigns are symbols formed by running together two characters into one without the
intercharacter space to make an abbreviation for the most common procedural signals. Usually
written with a BAR over the characters.

AR ----- End of message
AS ----- Stand by
BK ----- Invite receiving station to transmit
BT ----- Pause; Break For Text
KA ----- Beginning of message
KN ----- end of transmission
CL ----- Going off the air (clear)
SK ----- End of contact (sent before call)
R ------ All received OK
VE ----- Understood (VE)
K ------ Go, invite any station to transmit
KN ----- invite a specific station to transmit

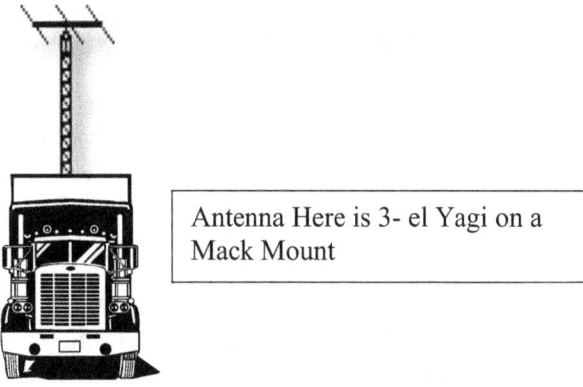

Antenna Here is 3- el Yagi on a Mack Mount

A1-4   CW BANDWIDTH AND CHARACTER SPACING

CW Bandwidth = wpm X 4 (e.g., 40 WPM = 160 Hz)
Character Spacing and Calculating Morse Code Speed

The word PARIS is the standard to determine CW code speed. Each dit is one element, each dah is three elements, intra-character spacing is one element, inter-character spacing is three elements and inter-word spacing is seven elements. The word PARIS is exactly 50 elements. Note that after each dit/dah of the letter P -- one element spacing is used except the last one. (Intra-Character). After the last dit of P is sent, 3 elements are added (Inter-Character). After the word PARIS - 7 elements are used.
Thus:
P
di da da di
1 1 3 1 3 1 1 (3) = 14 elements
A
di da
1 1 3 (3) = 8 elements
R
di da di
1 1 3 1 1 (3) = 10 elements
I
di di
1 1 1 (3) = 6 elements
S
di di di
1 1 1 1 1 [7] = 12 elements
Total = 50 elements
() = intercharacter
[] = interword

If you send PARIS 5 times in a minute (5WPM) you have sent 250 elements (using correct spacing). 250 elements into 60 seconds per minute = 240 milliseconds per element.

13 words-per-minute is one element every 92.31 milliseconds.

The Farnsworth method sends the dits and dahs and intra-character spacing at a higher speed, then increasing the inter-character and inter-word spacing to slow the sending speed down to the overall speed. For example, to send at 5 wpm with 13 wpm characters in Farnsworth method, the dits and intra-character spacing would be 92.3 milliseconds, the dah would be 276.9 milliseconds, the inter-character spacing would be 1.443 seconds and inter-word spacing would be 3.367 seconds.

Weighting is the ratio of dash length to dot length and can be changed for faster sending and ease of copy. One scheme is auto-weighting which increases the weighing as speed increases. In other keyers the weighting can be set as a percentage. Best experiment with your preferred weighting both for yourself and the opinions of others.

A1-5   IAMBIC KEYING

Iamb dates back to ancient Greek poetry. Iamb is a term from literature meaning a two-syllable rhythm. An "iamb" consists of two syllables. It may be two separate words, one word with two syllables, or even a portion of a multi-syllable word. But the important thing is that each of the two syllables has a different "accent." Sometimes the accent is called a SHORT/ LONG. That means, the first word comes quickly off the mouth when spoken; the second word actually sounds for a longer time. So somewhere along the line someone dubbed the squeeze key method of sending Morse characters as Iambic, perhaps because of the short/long (dit/dah) aspect.

Overall, single lever paddles take more mechanical motion to send characters than an iambic paddle. If you are just starting with an electronic keyer, most top CW ops will recommend an Iambic paddle. If you are used to slapping a key with quite a bit of force, you'll have to ease up on an Iambic paddle. Just a light touch is all it takes to work an Iambic paddle. If the paddle moves, then you are using too much force.

Be aware there are two different modes of Iambic operation – type A and type B. When a squeeze is released during an element (dot or dash), type "B" adds the opposite element. Type "A" just finishes the element in progress and does not produce a following alternate element. For example, in Type "A" Iambic, a squeeze release during the "dah" in the letter A will produce "dit dah" (A). In Type "B" Iambic, a squeeze release during the "dah" in the letter A will produce "dit dah dit" (R). That is if the left paddle is still depressed at the half-way point of the dah.

Usually, configure so your thumb sends the dits, wire accordingly for right or left handed operation. Some folks connect it in reverse and it is really a personal preference, but you may be in for a surprise if you operate some else's station. If you do encounter a "backwards" paddle, a quick fix that works on some paddles is to turn them upside down! I did this at our last Field Day rather than rewire temporally, however a lefty was right at home with it rightside up (;-)

Adjust the contact spacing for narrow spacing, as too wide a setting may cause problems.

Check that the paddle holds the adjustment and stops sending when you release pressure on the paddles. If there is a tension adjustment, adjust this for the minimum tension consistent with the contacts staying open. Close spacing and minimum tension will result in increased speed.

Engage the left paddle for a string of dits. Now engage the right paddle for a string of dahs. Next, press both paddles simultaneously. Use a squeezing motion with both fingers. This will generate a series of dits and dahs. For type B Iambic operation, note that when sending the letter "A" if you engage the left paddle too long – the keyer has a mind of it's own and sends "R". That is if the left paddle is still depressed at the half-way point of the dah. The first paddle contacted will determine whether a dot or dash occurs first.

The internal memory of the keyer automatically detects a long paddle press and after the dah has finished, the keyer will send an additional dit even though you have released pressure on the left paddle. The keyer will do the same thing for the opposite combination. From here lots of practice will get you used to the Iambic method, learning when the beast self-completes and how to allow spacing properly. Expert ops can send continuously without leaving their fingers off the paddles, so perhaps it is best described as touch-squeeze keying technique.

After a while the finger-brain activity becomes second nature and it will be as easy as single paddle or straight key sending, but a lot less tiring and it is probably the fastest keying method.

A1-6   ADJUSTING PADDLES AND STRAIGHT KEYS

For straight keys – see URL: http://www.mtechnologies.com/misc/keyadj.htm

For paddles: Since adjustments can be interactive, start by loosening all adjustments.

Adjust the bearing tension until you can just barely feel a bit of friction as you move the levers.

Adjust the contact spacing to about the thickness of a dime or less. The spacing can be different for the two contacts, some operators prefer a greater gap on the dash paddle. Some use a King of Spades card for the thickness.

Adjust the arm(s) tension for the minimum amount of tension that will still allow you to feel that you are in control of the paddle.

Close spacing and minimum tension will result in increased speed.

APPENDIX A2   CW ABBREVIATIONS

AA - All after AB - All before ABT - About ADEE - Addressee ADR - Address ADS - Address AGN - Again AM - Amplitude Modulation ANI - Any ANS - Answer ANT - Antenna BCI - Broadcast Interference BCL - Broadcast Listener BCNU - Be seeing you BD - Bad BK - Break, Break in BN - All between; Been BT - Separation (break) between addr & text; between txt & signature BTH - Both BTR - Better BUG - Semi-Automatic key BURO - Bureau B4 - Before	C -Yes, Correct CB - CallBook CBA - Callbook Address CFM - Confirm; I confirm CK - Check CKT - Circuit CL - I am closing my station; Call CLBK - Callbook CLD - Called CLG - Calling CMG - Coming CNT - Can't CONDX - Conditions CPI - Copy CQ - Calling any station CRD - Card CS - Call Sign CU - See You CUAGN - See You Again CUD- Could CUL - See You later CUM - Come CUZ - Because CW - Continuous wave	DA - Day DE - From, This Is DIFF - Difference DLD - Delivered DLVD - Delivered DN - Down DR - Dear DWN - Down DX - Distance EL - Element ES - And ENUF - Enough EU - Europe EVE - Evening FB - Fine Business, excellent FER - For FM - Frequency Modulation: From FONE - Phone FQ - Frequency Freq - Frequency FWD - Forward
GA - Go ahead; Good Afternoon GB - Good bye, God Bless GD - Good, Good Day GE - Good Evening GESS - Guess GG - Going GLD - Glad GM - Good morning GN - Good night GND - Ground GP - Ground Plane GS - Green Stamp GUD - Good GV - Give GVG - Giving	HH - Error in sending HI -The telegraph laugh; High HPE - Hope HQ - Headquarters HR - Here; Hear, Hour HRD - Heard HRS - Hours HRD - Heard HV - Have HVG - Having HVY - Heavy HW - How, How Copy?	II-- I Repeat IMI - Repeat, Say Again INFO - Info JA - Japanese Station K - Invitation To Transmit KLIX - Key Clicks LID - A poor operator LNG - Long LP - Long Path LSN - Listen LTR - Later; letter LV - Leave LVG -Leaving LW - Long Wire, Long Wave
MA - Milliampers MGR - Manager MI - My MILL - Typewriter MILS - Milliampers MNI - Many MOM - Moment MSG - Message; Prefix to radiogram MULT - Multiplier	N - No, Negative, Incorrect, No More N - Nine (as in Signal Report) NCS - Net Control Station ND - Nothing Doing NIL - Nothing; I have nothing for you NM - No more NR - Number, Near NW -Now; I resume transmission	OB - Old boy OC - Old chap OK - Correct OM - Old man OP - Operator OPR - Operator OT - Old timer; Old top OW - Old Woman

APPENDIX A2   CW   ABBREVIATIONS (Continued)

PBL - Preamble PKG - Package PSE - Please PT - Point PWR - Power PX - Press, Prefix	R - Received as transmitted; Are; Decimal Point RC - Ragchew RCD - Received RCVR - Receiver RE - Concerning; Regarding REF - Refer to; Referring to; Reference RFI - Radio frequency interference RIG - Station equipment RPT - Repeat, Report RTTY - Radio teletype RST - Readability, strength, tone RX - Receive, Receiver	SA - Say SASE - Self-addressed, stamped envelope SED - Said SAE - Self-Addressed Envelope SEZ - Says SGD - Signed SHUD - Should SIG - Signature; Signal SINE - Operator's personal initials or nickname SK - Silent Key SKED - Schedule SN - Soon SP - Short Path SRI - Sorry
SS - Sweepstakes SSB - Single Side Band STN - Station SUM -Some SVC - Service; Prefix to service message SWL - Short Wave Listener	T - Zero TEMP - Temperature TEST - Testing or Contest TFC - Traffic TMW - Tomorrow TKS - Thanks TNX - Thanks TR - Transmit T/R - Transmit/Receive TRBL - Trouble TRIX - Tricks TRX - Transceiver TT - That TTS - That is TU - Thank you TVI - Television interference TX - Transmitter; Transmit TXT - Text	U - You UFB - Ultra Fine Business UNLIS - Unlicensed UR - Your; You're URL - Universal Resource Locator Address For a WebPage URS - Yours VERT - Vertical VFB - Very fine business VFO - Variable Frequency Oscillator VY - Very
W - Watts WA - Word after WATSA - What Say WB - Word before WD - Word WDS - Words WID - With WKD - Worked WKG - Working WL - Well; Will WPM - Words Per Minute WRD - Word WRK - Work WUD - Would WW - Would WX - Weather	XCVR -Transceiver XMAS - Christmas XMTR -Transmitter XTAL - Crystal XYL - Wife YF - Wife YL - Young lady YR - Year Z - Zulu Time 30 - I have no more to send	33 - Fondest Regards 55 - Best Success 73 - Best Regards 88 - Love and kisses 161 - 73+88=161 "161" first came about in FOC circles (First-Class CW Operators' Club, founded by Louis Varney G5RV a number of years ago). The essential meaning is "Best regards to you and your XYL".

COMPILED BY AC6V, 2001

APPENDIX A3

The RST SYSTEM

The RST System of Signal Reporting has been used for years (circa 1934) as a shorthand method of reporting Readability, Signal Strength and for CW; Tone (i.e., quality of the CW tone). For voice contacts only the R and S are used. The S component is usually not the same as your S-Meter reading as most S-Meters aren't calibrated to track the RST System. The RST is also reported on QSL Cards and must be filled in correctly -- e.g., a 569 report for a Voice Contact is invalid. Note that many DX operations and contest stations merely report 59(9) as a convenience to avoid having to log each of the real reports. A questionable practice but a fact of DXing/Contesting.

READABILTY	CW TONE
1 Unreadable	1 Sixty cycle a.c. or less, very rough and broad
2 Barely readable, copy occasional words	2 Very rough a.c. , very harsh and broad
3 Readable with considerable difficulty	3 Rough a.c. tone, rectified but not filtered
4 Readable with practically no difficulty	4 Rough note, some trace of filtering
5 Perfectly readable	5 Filtered rectified a.c. but strongly ripple-modulated
SIGNAL STRENGTH	6 Filtered tone, definite trace of ripple modulation
1 Faint signals, barely perceptible	7 Near pure tone, trace of ripple modulation
2 Very weak signals	8 Near perfect tone, slight trace of modulation
3 Weak signals	9 Perfect tone, no trace of ripple or modulation of any kind
4 Fair signals	
5 Fairly good signals	
6 Good signals	
7 Moderately strong signals	
8 Strong signals	
9 Extremely strong signals	

Infrequently used is the addition of a letter to the end of the 3 numbers. These are: X = the signal is rock steady like a crystal controlled signal; C = the signal is chirpy as the frequency varies slightly with keying; and K = the signal has key clicks. X is from the early days of radio when such steady signals were rare. Today most all signals could be given an X but it is hardly ever used. It is helpful to report a chirpy or clicky signal by using the C or K, e.g. 579C or 579K.

It is very common to send RST reports in abbreviated form, for example 599, is sent as 5NN. "N" in place of the number "9". Also another time saver is for the zero using a long "T". "T" is sent in place of the number zero as in "POWER HR IS 3TT WATTS". Although there is a number code for all numbers, the abbreviated N and T codes are the most common ones. Also CW stations sometimes report their zones as "A4" or "A5" instead of sending "14" or "15".

1 = A,   2 = U,   3 = V,   4 = 4,   5 = E,   6 = 6,   7 = B,   8 = D,   9 = N,   0 = T

APPENDIX A4 DXER PHONETICS

NATO/ITU	DXer Phonetics
Alpha	America, Amsterdam
Bravo	Boston, Baltimore, Brazil
Charlie	Canada, Columbia, Chile
Delta	Denmark
Echo	England, Egypt
Foxtrot	France, Finland
Golf	Germany, Guatemala
Hotel	Honolulu, Hawaii
India	Italy
Juliet	Japan
Kilo	Kentucky, Kilowatt, King
Lima	London, Lima, Luxembourg
Mike	Mexico, Montreal
November	Norway, Nicaragua
Oscar	Ontario, Ocean,
Papa	Portugal, Pacific
Quebec	Queen
Romeo	Radio, Romania, Russia
Sierra	Santiago, Spain, Sweden
Tango	Tokyo, Texas
Uniform	United, Uruguay
Victor	Victoria, Venezuela
Whiskey	Washington
X-Ray	X-Ray
Yankee	Yokohama
Zulu	Zanzibar

RAQUEL SQUELCH   FRANKLY MY DEAR I DON'T QSL

Appendix A5. Q-Signals

Q-Sig	Message
QRA	What is the name of your station? The name of my station is___.
QBM	Has ... sent a message for me? Here is the message sent by ... at ... hours.
QRG	Will you tell me my exact frequency? Your exact frequency is ___ kHz.
QRH	Does my frequency vary? Your frequency varies.
QRI	How is the tone of my transmission? The tone of your transmission is ___ (1-Good, 2-Variable, 3-Bad.)
QRJ	Are you receiving me badly? I cannot receive you, your signal is too weak.
QRK	What is the intelligibility of my signals? The intelligibility of your signals is ___ (1-Bad, 2-Poor, 3-Fair, 4-Good, 5-Excellent.)
QRL	Are you busy? I am busy, please do not interfere
QRM	Is my transmission being interfered with? Your transmission is being interfered with ___ (1-Nil, 2-Slightly, 3-Moderately, 4-Severly, 5-Extremely.)
QRN	Are you troubled by static? I am troubled by static ___ (1-5 as under QRM.)
QRO	Shall I increase power? Increase power.
QRP	Shall I decrease power? Decrease power.
QRQ	Shall I send faster? Send faster (___ WPM.)
QRS	Shall I send more slowly? Send more slowly (___ WPM.)
QRT	Shall I stop sending? Stop sending.
QRU	Have you anything for me? I have nothing for you.
QRV	Are you ready? I am ready.
QRX	When will you call me again? I will call you again at ___ hours.
QRZ	Who is calling me? You are being called by ___.
QSA	What is the strength of my signals? The strength of your signals is ___ (1-Scarcely perceptible, 2-Weak, 3-Fairly Good, 4-Good, 5-Very Good.)
QSB	Are my signals fading? Your signals are fading.
QSK	Can you hear me between you signals and if so can I break in on your transmission? I can hear you between my signals, break in on my transmission.

QSL	Can you acknowledge receipt? I am acknowledging receipt.
QSN	Did you hear me on ___ kHz? I did hear you on ___ kHz.
QSO	Can you communicate with ___ direct or by relay? I can communicate with ___ direct (or by relay through ___.)
QSP	Will you relay to ___? I will relay to ___.
QSU	Shall I send or reply on this frequency? Send a series of Vs on this frequency.
QSY	Shall I change to another frequency? Change to another frequency.
QSZ	Shall I send each word or group more than once? Send each word or group twice (or ___ times.)
QTC	How many messages have you to send? I have ___ messages for you.
QTH	What is your location? My location is ___.

Note that Q signals can take the form of a question when followed by a question mark.

Compiled By AC6V, 2001

APPENDICES A6 AND A7 CQ and ITU Zones

There are two different zone allocations for Amateur Radio, the CQ Magazine zones and the zones from the ITU, Some contests use the CQ zone as a part of the report while others use the ITU zone. Check the contest rules. CQ Magazine offers awards for worked all zones (WAZ) on the various bands and modes

CQ ZONES AND ITU ZONE MAPS

These are shown in Appendix A6 and A7 on the next page. The maps are courtesy of VE3NEA Alex Shovkoplyas, author of the DX Atlas program (highly recommended for Dxers) See URL: http://www.dxatlas.com

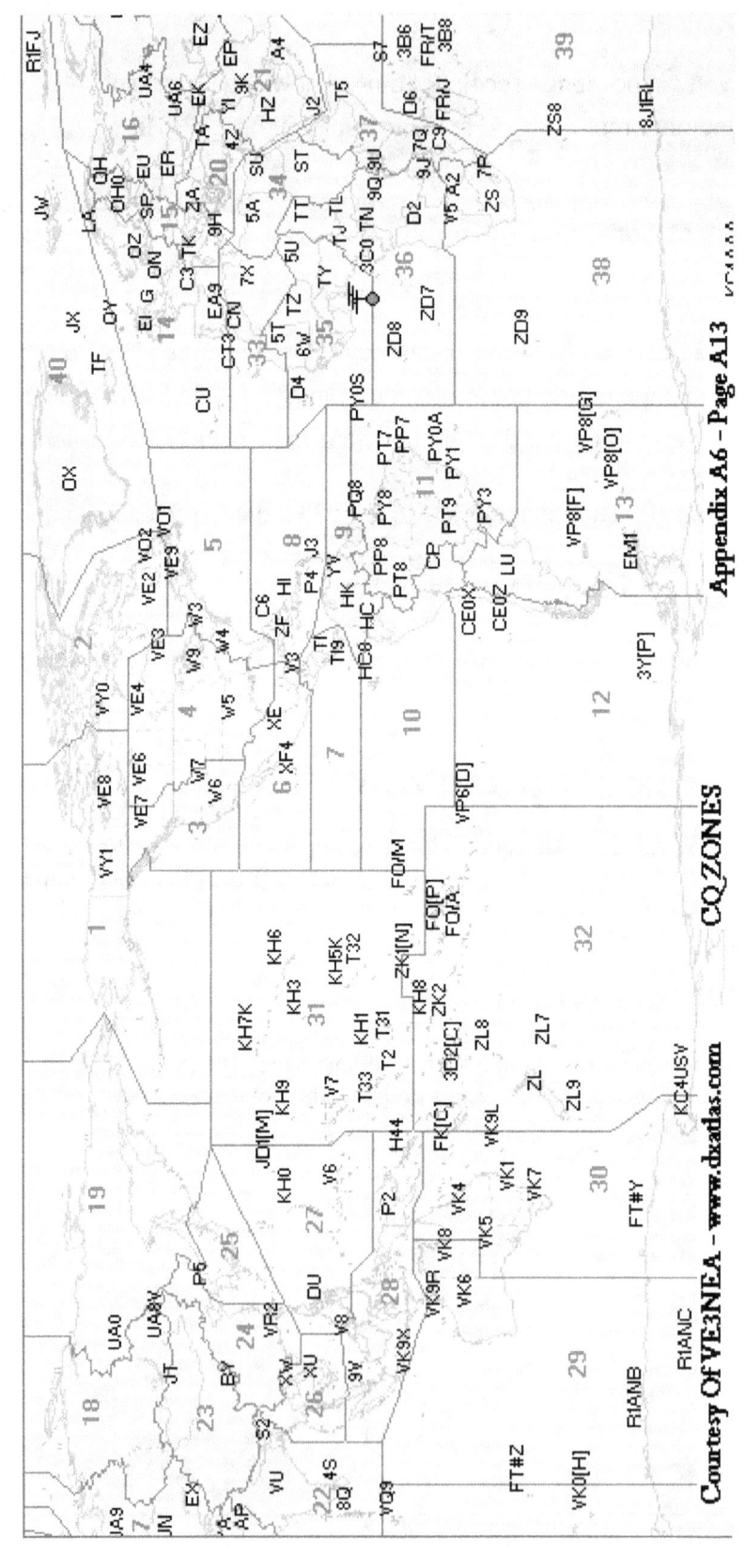

CQ ZONES

Appendix A6 – Page A13

Courtesy Of VE3NEA – www.dxatlas.com

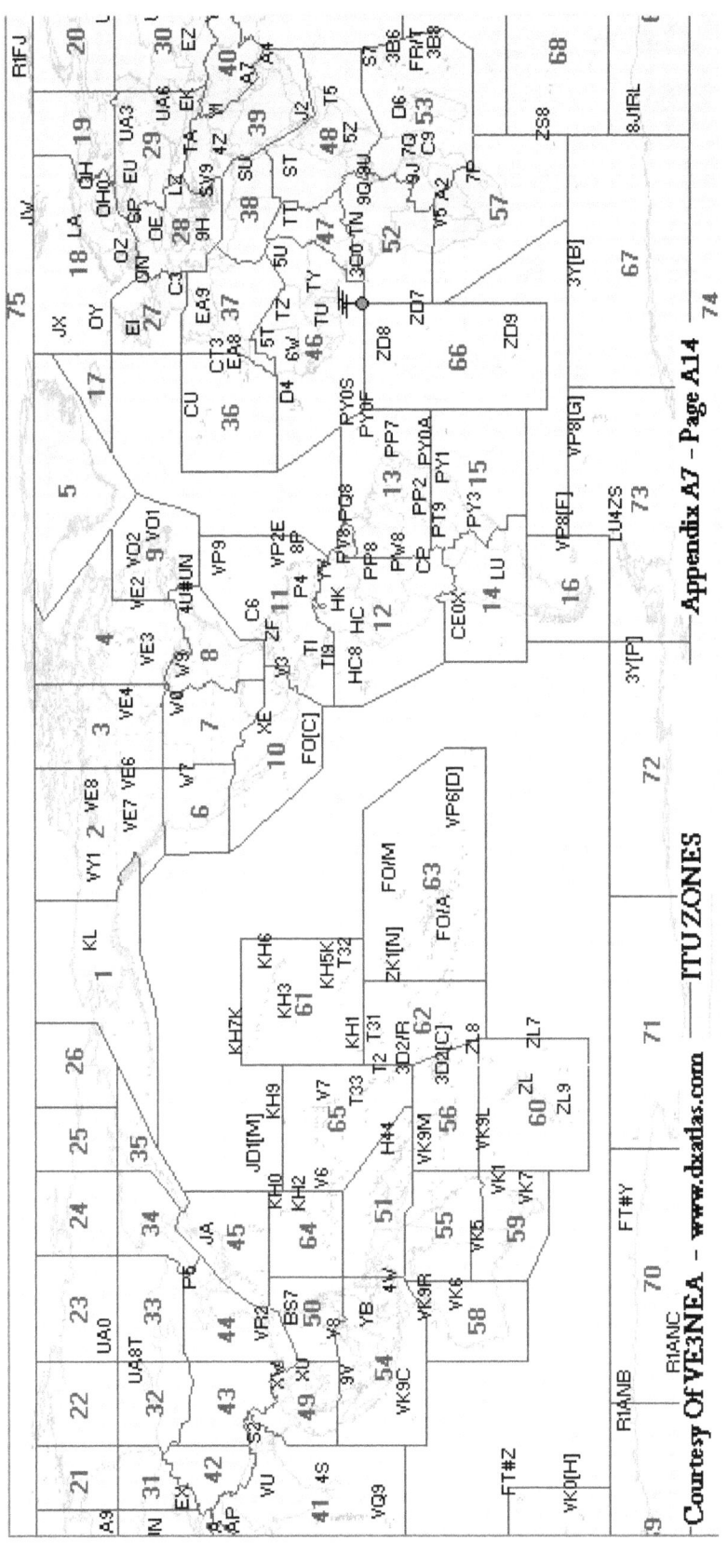

Courtesy Of VE3NEA – www.dxatlas.com —— ITU ZONES —— Appendix A7 – Page A14

APPENDIX A8 WORKED ALL CQ ZONES

Zone	MIX	SSB	CW	Data	160	80	40	30	20	17	15	12	10	6
1														
2														
3														
4														
5														
6														
7														
8														
9														
10														
11														
12														
13														
14														
15														
16														
17														
18														
19														
20														
21														
22														
23														
24														
25														
26														
27														
28														
29														
30														
31														
32														
33														
34														
35														
36														
37														
38														
39														
40														

APPENDIX A9

W0 (CO) Colorado ITU 7 Zone CQ 4
W0 (IA) Iowa ITU 7 CQ 4
W0 (KS) Kansas ITU 7 CQ 4
W0 (MN) Minnesota ITU 7 CQ 4
W0 (MO) Missouri ITU 7 CQ 4
W0 (NE) Nebraska ITU 7 CQ 4
W0 (ND) North Dakota ITU 7 CQ 4
W0 (SD) South Dakota ITU 7 CQ 4
W1 (CT) Connecticut ITU 08 CQ 5
W1 (ME) Maine ITU 08 CQ 5
W1 (MA) Massachusetts ITU 08 CQ 5
W1 (NH) New Hampshire ITU 08 CQ 5
W1 (RI) Rhode Island ITU 08 CQ 5
W1 (VT) Vermont ITU 08 CQ 5
W2 (NJ) New Jersey ITU 08 CQ 5
W2 (NY) New York ITU 08 CQ 5
W3 D.C. ITU 08 CQ 5
W3 (DE) Delaware ITU 08 CQ 5

W3 (MD) Maryland ITU 08 CQ 5
W3 (PA) Pennsylvania ITU 08 CQ 5
W4 (AL) Alabama ITU 08 CQ 4
W4 (FL) Florida ITU 08 CQ 5
W4 (GA) Georgia ITU 08 CQ 5
W4 (KY) Kentucky ITU 08 CQ 4
W4 (NC) North Carolina ITU 08 CQ 5
W4 (SC) South Carolina ITU 08 CQ 5
W4 (TN) Tennessee ITU 08 CQ 4
W4 (VA) Virginia ITU 08 CQ 5
W5 (AR) Arkansas ITU 7 CQ 4
W5 (LA) Louisiana ITU 7 CQ 4
W5 (MS) Mississippi ITU 08 CQ 4
W5 (NM) New Mexico ITU 7 CQ 4
W5 (OK) Oklahoma ITU 7 CQ 4
W5 (TX) Texas ITU 7 CQ 4
W6 (CA) California ITU 6 CQ 3
W7 (AZ) Arizona ITU 6 CQ 3
W7 (ID) Idaho ITU 6 CQ 3

W7 (MT) Montana ITU 6 (excluding Montana east of 110W) (ITU 7 Montana east of 110W), Both are CQ 4

W7 (NV) Nevada ITU 6 CQ 3
W7 (OR) Oregon ITU 6 CQ 3
W7 (UT) Utah ITU 6 CQ 3
W7 (WA) Washington ITU 6 CQ 3

W7 (WY) Wyoming ITU 6 (excluding Wyoming east of 110W) (ITU 7 Wyoming east of 110W), Both are CQ 4

W8 (MI) Michigan ITU 08 CQ 4
W8 (OH) Ohio ITU 08 CQ 4
W8 (WV) West Virginia ITU 08 CQ 5
W9 (IL) Illinois ITU 08 CQ 4
W9 (IN) Indiana ITU 08 CQ 4
W9 (WI) Wisconsin ITU 08 CQ 4

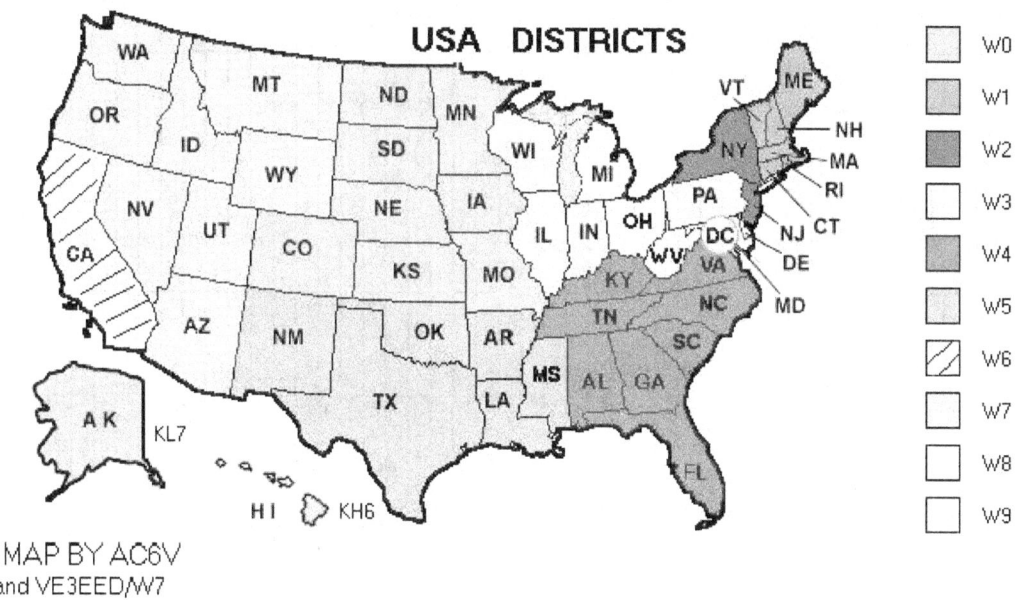

MAP BY AC6V
and VE3EED/W7

1A Mil Or Malta (15)	4U1WB Wrld Bnk (05)	8Q Maldive Is (22)	AO9 Ceuta-Melilla (33)	DS, DT So Korea (25)	GC Wales (Clubs) (14)	HZ Saudi Arabia (21)
1B No Cyprus* (20)	4V Haiti (08)	8R Guyana (09)	AP- AS Pakistan (21)	DU to DZ Philippines (27)	GD Isle of Man (14)	I Italy (15)
1B Chechnya*	4W East Timor (28)	8S Sweden (14)	AT thru AW India (22)	E2 Thailand (26)	GH Jersey (Clubs) (14)	T9, IW9 Sicily (15)
1S Spratly Is (26)	4X Israel (20)	8T thru 8Y India (22)	AT4 Andaman Is (26)	E3 Eritrea (37)	GI Northern Ireland (14)	ISO Sardinia (15)
1SL Sealand	4Y Int Civil Aviation	8Z Saudi Arabia (21)	AT7 Laccadive Is (22)	E4 Palestine (20)	GJ Jersey (14)	J2 Djibouti (37)
2E England (14)	4Z Israel (20)	9A Croatia (15)	AU4 Andaman Is (26)	E7 Bosnia/Herzegovina	GM Scotland (14)	J3 Grenada (08)
2D Isle Of Man (14)	5A Libya (34)	9B, 9C, 9D Iran (21)	AU7 Laccadive Is (22)	EA thru EH Spain (14)	GN Northern Ireland 914)	J4 Greece (20)
2I No Ireland (14)	5B Cyprus (20)	9E, 9F Ethiopia (37)	AV4 Andaman Is (26)	EA6&EA7 Balearic Is(14)	GP Guernsey(Clubs)(14)	J45 Dodecanese (20)
2J Jersey (14)	5C - 5G Morocco (33)	9G Ghana (35)	AV7 Laccadive Is (22)	EA8 Canary Is (33)	GS Scotland (Clubs)(14)	J49 Crete (20)
2M Scotland (14)	5H, 5I Tanzania (37)	9H Malta (15)	AW4 Andaman Is (26)	EA9 Ceuta (33)	GT Is Of Man (Clubs)(14)	J5 Guinea Bissau (35)
2U Guernsey (14)	5J, 5K Columbia (09)	9I, 9J Zambia (36)	AW7 Laccadive Is (32)	EI, EJ Ireland (14)	GU Guernsey (14)	J6 St Lucia (08)
2W Wales (14)	5L, 5M Liberia (35)	9K Kuwait (21)	AX Australia (29,30)	EK Armenia (21)	GW Wales (14)	J7 Dominica (08)
3A Monaco (14)	5N, 5O Nigeria (35)	9L Sierra Leone (35)	AY, AZ Argentina (13)	EL Liberia (35)	GX England (Clubs) (14)	J8 St Vincent (08)
3B6 Agalega (39)	5P, 5Q Denmark (14)	9M0 Spratly Is (26)	BA - BU, BV - BZ China	EM thru EO Ukraine (16)	GZ Shetland Isl (14)	JA thru JS Japan (25)
3B7 Agalega (39)	5R, 5S Malagasy (39)	9M2 West Malaysia (28)	BV Taiwan (24)	EP, EQ Iran (21)	H2 Cyprus (20)	JD Ogasawara (27)
3B8 Mauritius (39)	5T Mauritania (35)	9M4 West Malaysia (28)	C2 Nauru (31)	ER Moldova (16)	H3 Panama (07)	JD Min Torishima (27)
3B9 Rodriguez (39)	5U Niger (35)	9M6 East Malaysia (28)	C3 Andorra (14)	ES Estonia (15)	H4 Solomon Is (28)	JT,JU,JV Mongolia (23)
3C Eq Guinea (36)	5V Togo (35)	9M8 East Malaysia (28)	C4 Cyprus (20)	ET Ethiopia (37)	H40 Temotu (28)	JW Svalbard Is (40)
3C0 Annobon (36)	5W West Samoa (32)	9N Nepal (22)	C5 Gambia (35)	EU,EV,EW Belarus (16)	H6, H7 Nicaragua (07)	JX Jan Mayen (40)
3D6 Swaziland (38)	5X Uganda (37)	9O thru 9T Zaire (36)	C6 Bahamas (08)	EX Kyrgyzstan (17)	H8, H9 Panama (07)	JY Jordan (20)
3DA0 Swaziland (38)	5Y, 5Z Kenya (37)	9U Burundi (36)	C7 World Meteorological	EY Tadjikistan (17)	HA Hungary (15)	JZ Indonesia (28)
3D2 Conway Reef (32)	6A, 6B Egypt (34)	9V Singapore (28)	C8, C9 Mozambique (37)	EZ Turkmenistan (17)	HB Switzerland (14)	KA-KZ USA See W's
3D2/C Conway Reef (32)	6C Syria (20)	9W W & E Malaysia (28)	CA thru CE Chile (12)	F France (14)	HB0 Liechtenstein (14)	KC4 Antarct Bryd (12)
3D2/F Fiji Islands (32)	6D - 6J Mexico (06)	9X Rwanda (36)	CF-CK Canada (1-5)	FG Guadeloupe (08)	HC, HD Ecuador (10)	KC4 McMurdo (30)
3D2/R Rotuma (32)	6K - 6N S. Korea (25)	9Y, 9Z Trinidad (09)	CL, CM Cuba (08)	FH Mayotte (39)	HC8, HD8 Galapagos(10)	KC4 Palmer (13)
3E, 3F Panama (07)	6O Somalia (37)	A2 Botswana (38)	CN Morocco (33)	FJ St Martin, St Barts (08)	HE Switzerland (14)	KG4 Guantanamo (08)
3G Chile (12)	6P - 6S Pakistan (21)	A3 Tonga (32)	CO Cuba (08)	FK Chesterfield Isle (32)	HF Poland (15)	LA thru LN Norway (14)
3H - 3U China (23, 24)	6T, 6U Sudan (34)	A4 Oman (21)	CP Bolivia (10)	FK New Caledonia (32)	HF0 So Shetland (13)	L2 - L9 Argentina (13)
3V Tunisia (33)	6V, 6W Senegal (35)	A5 Bhutan (22)	CQ - CU Portugal (14)	FM Martinique (08)	HF0 Antarctica	LO - LW Argentina (13)
3W Vietnam (26)	6X Malagasy Rep (39)	A6 Un Arab Emir (21)	CR3 & CR9 Madeira (33)	FO Australs (32)	HG Hungary (15)	LX Luxembourg (14)
3X Guinea (35)	6Y Jamaica (08)	A7 Qatar (21)	CS3 & CS9 Madeira (33)	FO Clipperton (07)	HH Haiti (08)	LY Lithuania (15)
3Y/B Bouvet (38)	6Z Liberia (35)	A8 Liberia (35)	CT3 Madeira Island (33)	FO Tahiti (32)	HI Dominican Rep (08)	LZ Bulgaria (20)
3Y/P Peter Is (12)	7A - 7I Indonesia (28)	A9 Bahrain (21)	CT4 Portugal (14)	FO Marquesas Is (31)	HJ, HK Columbia (09)	M England (14)
3Z Poland (15)	7J thru 7N Japan (25)	AAA-ALZ USA See W's	CT9 Madeira Isl.(33)	FP St Pierre Miq (05)	HK0/M Malpelo Is (10)	MC Wales (Clubs) (14)
4A - 4C Mexico (06)	7O Yemen Rep (21)	AM, AN, AO Spain (14)	CU Azores Isl. (14)	FR Glorioso (39)	HK0 San Andres (07)	MD Isle of Man (14)
4D - 4I Philippines (27)	7P Lesotho (38)	AM6 Balearic Is (14)	CV,CW,CX Uruguay(13)	FR Juan De Nova (39)	HL South Korea (25)	MH Jersey (Clubs) (14)
4J, 4K Azerbaijan (21)	7Q Malawi (37)	AM8 Canary Is (33)	CY,CZ Canada (1-5)	FR Reunion (39)	HM North Korea (25)	MI Northern Ireland (14)
4L Georgia (21)	7R Algeria (33)	AM9 Ceuta-Melilla (33)	CY0 Sable Is (05)	FR Tromelin (39)	HN Iraq (21)	MJ Jersey (14)
4M Venezuela (09)	7S Sweden (14)	AN Spain (14)	CY9 St Paul Is (05)	FS St Martin (08)	HO, HP Panama (07)	MM Scotland (14)
4O Montenegro (15)	7T - 7Y Algeria (33)	AN6 Balearic Is (14)	D2, D3 Angola (36)	FT0W Crozet (39)	HQ, HR Honduras (07)	MN No Ireland (Clubs) (14)
4P-4S Sri Lanka (22)	7Z Saudi Arabia (21)	AN8 Canary Is (33)	D4 Cape Verde (35)	FT0X Kerguelen (39)	HS Thailand (26)	MP Guernsey (Clubs) (14)
4T Peru (10)	8A - 8I Indonesia (28)	AN9 Ceuta-Melilla (33)	D5 Liberia (35)	FT0Z Amsterdam (39)	HT Nicaragua (07)	MS Scotland (Clubs) (14)
4U1ITU Geneva (14)	8J - 8N Japan (25)	AO Spain (14)	D6 Comoros (39)	FW Wallis (32)	HU El Salvador (07)	MT Isle O Man (Clubs) (14)
4U1UN NY NY (05)	8O Botswana (38)	AO6 Balearic Is (14)	D7 to D9 So Korea (25)	FY Fr Guiana (09)	HV Vatican City (15)	MU Guernsey (14)
4U1VIC Vienna (15)	8P Barbados (08)	AO8 Canary Is (33)	DA-DR Germany (14)	G England (14)	HW - HY France (14)	MW Wales (14)

MX England (Clubs) (14)	SN thru SR Poland (15)	UR-UZ Ukraine (16)	WA-WZ USA (Zone 3,4, 5)	W8 (OH) Ohio (04)	YK Syria (20)
MZ Shetland Islands (13)	SS Egypt (34)	V2 Antigua (08)	W0 (CO) Colorado (04)	W8 (WV) West Virginia (05)	YL Latvia (15)
NA-NZ USA See W's	ST Sudan (34)	V3 Belize (07)	W0 (IA) Iowa (04)	W9 (IL) Illinois (04)	YM Turkey (20)
OA, OB, OC Peru (10)	SU Egypt (34)	V4 St Kitts (08)	W0 (KS) Kansas (04)	W9 (IN) Indiana (04)	YN Nicaragua (07)
OD Lebanon (20)	SV thru SZ Greece (20)	V5 Namibia (38)	W0 (MN) Minnesota (04)	W9 (WI) Wisconsin (04)	YO thru YR Romania (20)
OE Austria (15)	SV/A Mount Athos (20)	V6 Fed Micronesia (27)	W0 (MO) Missouri (04)	WH0 Mariana Is (27)	YS El Salvador (07)
OF thru OJ Finland (15)	SV5 Dodecanese (20)	V7 Marshall Is (31)	W0 (NE) Nebraska (04)	WH1 Baker Howland (31)	YT Serbia (15)
OH0 Aland Is (15)	SV9 Crete (20)	V8 Brunei (28)	W0 (ND) North Dakota (04)	WH2 Guam (27)	YU Serbia (15)
OH0M Market Reef (15)	T2 Tuvalu (31)	VA-VG Canada (1-5)	W0 (SD) South Dakota (04)	WH3 Johnston Is (31)	YV thru YY Venezuela (09)
OJ0 Market Reef (15)	T30 West Kiribati (31)	VH-VN Australia	W1 (CT) Connecticut (05)	WH4 Midway Is (31)	YV0 Aves Is (08)
OK,OL Czech Republic (15)	T31 Central Kiribati (31)	VK9/C Cocos Keeling (29)	W1 (ME) Maine (05)	WH5 Palmyra Is (31)	YZ Returned To ITU
OM Slovakia (15)	T32 East Kiribati (31)	VK9/H Lord Howe (30)	W1 (MA) Massachusetts (05)	WH5K Kingman Reef (31)	Z2 Zimbabwe (38)
ON thru OT Belgium (14)	T88 Belau (Palau)(KC6) (27)	VK9/M Mellish Reef (30)	W1 (NH) New Hampshire (05)	WH6 Hawaii (31)	Z3 Macedonia (15)
OU thru OZ Denmark (14)	T9 Bosnia Now E7 (15)	VK9/N Norfolk Is (32)	W1 (RI) Rhode Island (05)	WH7 Kure Is (31)	Z3 Macedonia (15)
OX Greenland (40)	TA Turkey (20)	VK9/W Willis Is (30)	W1 (VT) Vermont (05)	WH7K Kure Is (31)	ZB Gibraltar (14)
OY Faroe Is (14)	TD Guatemala (07)	VK9X Christmas Is (29)	W2 (NJ) New Jersey (05)	WH8 American Samoa (32)	ZC4 UK Sov Base (20)
P2 Papua New Guinea (28)	TF Iceland (40)	VK0 Macquarie Is (30)	W2 (NY) New York (05)	WH8 Swains Is (32)	ZD7 St Helena (36)
P3 Cyprus (20)	TG Guatemala (07)	VK0 Heard Is (39)	W3 (DE) Delaware (05)	WH9 Wake Is (31)	ZD8 Ascension Is (36)
P4 Aruba (09)	TH France (14)	VOA-VOZ Canada (01-05)	W3 (MD) Maryland (05)	WL Alaska (01)	ZD9 Tristan Da Cunha (38)
P5 thru P9 North Korea (25)	TI Costa Rica (07)	VO1 Newfoundland (05)	W3 (PA) Pennsylvania (05)	WP1 Navassa Is (08)	ZF Cayman Is (08)
PA thru PI Netherlands (14)	TI9 Cocos Is (07)	VO2 Labrador (05)	W4 (AL) Alabama (04)	WP2 Virgin Is (08)	ZK1/N No Cook Is--(02)
PJ0 - PJ4 Neth Antilles (09)	TJ Cameroon (36)	VP2E Anguilla (08)	W4 (FL) Florida (05)	WP3, WP4 Puerto Rico (08)	ZK1/S So Cook Is--(32)
PJ5 - PJ7 St Maarten (08)	TK Corsica (15)	VP2M Montserrat (08)	W4 (GA) Georgia (05)	WP5 Desecheo Is (08)	ZK2 Niue Is (32)
PJ9 Neth Antilles (09)	TL Central Africa Rep (36)	VP2V Brit Virg Isle (08)	W4 (KY) Kentucky (04)	XA thru XI Mexico (06)	ZK3 Tokelaus (31)
PK thru PO Indonesia (28)	TM Outside France	VP5 Turks Caicos (08)	W4 (NC) North Carolina (05)	XF Revilla Gigedo (06)	ZL New Zealand (32)
PP thru PY Brazil (11)	TN Congo (36)	VP6D Ducie Isle (32)	W4 (SC) South Carolina (05)	XJ thru XO Canada (01-05)	ZL5 Antarctica Scott
PZ Suriname (09)	TO Outside France	VP6 Pitcairn (old VR6) (32)	W4 (TN) Tennessee (04)	XP Denmark (14)	ZL7 Chatham Is (32)
RA1 to RZ1 European Russia	TR Gabon (36)	VP8/F Falkland Is (13)	W4 (VA) Virginia (05)	XP Greenland (40)	ZL8 Kermadec Is (32)
RA2 to RZ2 Baltic Kaliningrad	TS Tunisia (33)	VP8/G So Georgia (13)	W5 (AR) Arkansas (04)	XQ, XR Chile (12)	ZL9 Auckland Campbell (32)
RA3 to RZ3 Central Russia	TT Chad (36)	VP8/O So Orkney (13)	W5 (LA) Louisiana (04)	XQ0X San Felix (12)	ZL0 Overseas Visitors (32)
RA4 to RZ4 Eu Russia-Volga	TU Ivory Coast (35)	VP8/SA So Sandwich (13)	W5 (MS) Mississippi (04)	XQ0Y Easter Island (12)	ZM New Zealand (32)
RA5 to RZ5 (Reserved)	TV -TX France	VP8/SH So Shetland (13)	W5 (NM) New Mexico (04)	XQ0Z Juan Fernandez (12)	ZM7 Chatham Is (32)
RA6-RZ6 Eu-No Caucasus	TX Chesterfield Isle (32)	VP8 Antarctica	W5 (OK) Oklahoma (04)	XR Chile (12)	ZM8 Kermadec Is (32)
RA7 to RZ7 (Reserved)	TY Benin (35)	VP9 Bermuda (05)	W5 (TX) Texas (04)	XR0Y Easter Island (12)	ZM9 Auckland Campbell (32)
RA8-RZ8 Asiatic Russia E	TZ Mali (35)	VQ9 Chagos (39)	W6 (CA) California (03)	XR0Z Easter Island (12)	ZP Paraguay (11)
RA9-RZ9 As Russia-W. Siberia	UA0-UI0 See RA - RZ	VRA-VRZ China	W7 (AZ) Arizona (03)	XS China (23,24)	ZR thru ZU So Africa (38)
RA0-RZ0 As Russia-E. Siberia	UJ-UM Uzbekistan (17)	VR2 Hong Kong (24)	W7 (ID) Idaho (03)	XT Burkina Faso (35)	ZS2 Marion Is (38)
S0 Western Sahara (33)	UN-UQ Kazakhstan (17)	VS6 Hong Kong (24)	W7 (MT) Montana (04)	XU Kampuchea (26)	ZV thru ZZ Brazil (11)
S2, S3 Bangladesh (22)		VT-VW India (22)	W7 (NV) Nevada (03)	XV Vietnam (26)	ZX0F Fern De Noronha (11)
S5 Slovenia (15)		VU7 Laccadive (22)	W7 (OR) Oregon (03)	XW Laos (26)	ZX0S Peter Paul Rocks (11)
S6 Singapore		VU7 Andaman&Nicbar (26)	W7 (UT) Utah (03)	XX Macao (24)	ZX0T, ZY0T Trindade (11)
S7 Seychelles (39)		VX, VY Canada (01-05)	W7 (WA) Washington (03)	XY, XZ Burma (26)	ZY0F Fern De Noronha (11)
S8 South Africa (38)		VY1 Yukon (01)	W7 (WY) Wyoming (04)	YA Afghanistan (21)	ZY0S Peter Paul Rocks (11)
S9 Sao Tome (36)		VY2, VY9 Prince Edward Is (05)	W8 (MI) Michigan (04)	YB thru YH Indonesia (28)	ZZ0F Fern De Noronha (11)
SA thru SM Sweden (14)		VY0 Nunavut		YI Iraq (21)	ZZ0S Peter Paul Rocks (11)
		VZ Australia (30)		YJ Vanuatu (32)	ZZ0T Trindade (11)

Prefix	Entity	CQ	Mix	SSB	CW	Data	80	40	30	20	17	15	12	10	
1A	SMO Malta	15					\	\	\	\	\	\	\	\	\
1S	Spratly Is	26					\	\	\	\	\	\	\	\	\
3A	Monaco	14					\	\	\	\	\	\	\	\	\
3B6	Agalega & St Brandon	39					\	\	\	\	\	\	\	\	\
3B8	Mauritius	39					\	\	\	\	\	\	\	\	\
3B9	Rodriguez Is	39					\	\	\	\	\	\	\	\	\
3C	Equatorial Guinea	36					\	\	\	\	\	\	\	\	\
3C0	Pagalu	36					\	\	\	\	\	\	\	\	\
3D2	Conway Reef	32					\	\	\	\	\	\	\	\	\
3D2	Fiji	32					\	\	\	\	\	\	\	\	\
3D2	Rotuma	32					\	\	\	\	\	\	\	\	\
3DA	Swaziland	38					\	\	\	\	\	\	\	\	\
3V	Tunisia	33					\	\	\	\	\	\	\	\	\
3W	Vietnam	26					\	\	\	\	\	\	\	\	\
3X	Guinea	35					\	\	\	\	\	\	\	\	\
3Y	Bouvet	38					\	\	\	\	\	\	\	\	\
3Y	Peter I	12					\	\	\	\	\	\	\	\	\
4J	Azerbaijan	21					\	\	\	\	\	\	\	\	\
4L	Georgia	21					\	\	\	\	\	\	\	\	\
4O	Montenegro	15					\	\	\	\	\	\	\	\	\
4S	Sri Lanka	22					\	\	\	\	\	\	\	\	\
4U	ITU Geneva	14					\	\	\	\	\	\	\	\	\
4U	UN HQ	5					\	\	\	\	\	\	\	\	\
4W	East Timor	28					\	\	\	\	\	\	\	\	\
4X	Israel	20					\	\	\	\	\	\	\	\	\
5A	Libya	34					\	\	\	\	\	\	\	\	\
5B	Cyprus	20					\	\	\	\	\	\	\	\	\
5H	Tanzania	37					\	\	\	\	\	\	\	\	\
5N	Nigeria	35					\	\	\	\	\	\	\	\	\
5R	Madagascar	39					\	\	\	\	\	\	\	\	\
5T	Mauritania	35					\	\	\	\	\	\	\	\	\
5U	Niger	35					\	\	\	\	\	\	\	\	\
5V	Togo	35					\	\	\	\	\	\	\	\	\
5W	Samoa	32					\	\	\	\	\	\	\	\	\
5X	Uganda	37					\	\	\	\	\	\	\	\	\
5Z	Kenya	37					\	\	\	\	\	\	\	\	\
6W	Senegal	35					\	\	\	\	\	\	\	\	\
6Y	Jamaica	8					\	\	\	\	\	\	\	\	\
7O	Yemen	21					\	\	\	\	\	\	\	\	\
7P	Lesotho	38					\	\	\	\	\	\	\	\	\
7Q	Malawi	37					\	\	\	\	\	\	\	\	\
7X	Algeria	33					\	\	\	\	\	\	\	\	\
8P	Barbados	8					\	\	\	\	\	\	\	\	\
8Q	Maldives	22					\	\	\	\	\	\	\	\	\
8R	Guyana	9					\	\	\	\	\	\	\	\	\
9A	Croatia	15					\	\	\	\	\	\	\	\	\
9G	Ghana	35					\	\	\	\	\	\	\	\	\
9H	Malta	15					\	\	\	\	\	\	\	\	\
9J	Zambia	36					\	\	\	\	\	\	\	\	\
9K	Kuwait	21					\	\	\	\	\	\	\	\	\
9L	Sierra Leone	35					\	\	\	\	\	\	\	\	\
9M2	Malaysia	28					\	\	\	\	\	\	\	\	\
9M6	East Malaysia	28					\	\	\	\	\	\	\	\	\
9N	Nepal	22					\	\	\	\	\	\	\	\	\
9Q	Dem Rep Congo	36					\	\	\	\	\	\	\	\	\
9U	Burundi	36					\	\	\	\	\	\	\	\	\

A-19 This is not a prefix list, See Appendix A12 for oddball prefixes

Prefix	Entity	CQ	Mix	SSB	CW	Data	80	40	30	20	17	15	12	10	
9V	Singapore	28					\	\	\	\	\	\	\	\	\
9X	Rwanda	36					\	\	\	\	\	\	\	\	\
9Y	Trinidad	9					\	\	\	\	\	\	\	\	\
A2	Botswana	38					\	\	\	\	\	\	\	\	\
A3	Tonga	32					\	\	\	\	\	\	\	\	\
A4	Oman	21					\	\	\	\	\	\	\	\	\
A5	Bhutan	22					\	\	\	\	\	\	\	\	\
A6	United Arab E	21					\	\	\	\	\	\	\	\	\
A7	Qatar	21					\	\	\	\	\	\	\	\	\
A9	Bahrain	21					\	\	\	\	\	\	\	\	\
AP	Pakistan	21					\	\	\	\	\	\	\	\	\
BS7	Scarborough	27					\	\	\	\	\	\	\	\	\
BV	Taiwan	24					\	\	\	\	\	\	\	\	\
BV9	Pratas Is	24					\	\	\	\	\	\	\	\	\
BY	China	24					\	\	\	\	\	\	\	\	\
C2	Nauru	31					\	\	\	\	\	\	\	\	\
C3	Andorra	14					\	\	\	\	\	\	\	\	\
C5	Gambia	35					\	\	\	\	\	\	\	\	\
C6	Bahamas	8					\	\	\	\	\	\	\	\	\
C9	Mozambique	37					\	\	\	\	\	\	\	\	\
CE	Chile	12					\	\	\	\	\	\	\	\	\
CE0X	San Felix	12					\	\	\	\	\	\	\	\	\
CE0Y	Easter Is	12					\	\	\	\	\	\	\	\	\
CE0Z	Juan Fernand	12					\	\	\	\	\	\	\	\	\
CE9	Antarctica						\	\	\	\	\	\	\	\	\
CN	Morocco	33					\	\	\	\	\	\	\	\	\
CO	Cuba	8					\	\	\	\	\	\	\	\	\
CP	Bolivia	10					\	\	\	\	\	\	\	\	\
CT	Portugal	14					\	\	\	\	\	\	\	\	\
CT3	Madeira Is	33					\	\	\	\	\	\	\	\	\
CU	Azores	14					\	\	\	\	\	\	\	\	\
CX	Uruguay	13					\	\	\	\	\	\	\	\	\
CY0	Sable Is	5					\	\	\	\	\	\	\	\	\
CY9	St Paul Is	5					\	\	\	\	\	\	\	\	\
D2	Angola	36					\	\	\	\	\	\	\	\	\
D4	Cape Verde	35					\	\	\	\	\	\	\	\	\
D6	Comoros	39					\	\	\	\	\	\	\	\	\
DL	Germany	14					\	\	\	\	\	\	\	\	\
DU	Philippines	27					\	\	\	\	\	\	\	\	\
E3	Eritrea	37					\	\	\	\	\	\	\	\	\
E4	Palestine	20					\	\	\	\	\	\	\	\	\
E7	Bosnia/Herzegovina	15					\	\	\	\	\	\	\	\	\
EA	Spain	14					\	\	\	\	\	\	\	\	\
EA6	Balearic Is	14					\	\	\	\	\	\	\	\	\
EA8	Canary Is	33					\	\	\	\	\	\	\	\	\
EA9	Ceuta & Melilla	33					\	\	\	\	\	\	\	\	\
EI	Ireland	14					\	\	\	\	\	\	\	\	\
EK	Armenia	21					\	\	\	\	\	\	\	\	\
EL	Liberia	35					\	\	\	\	\	\	\	\	\
EP	Iran	21					\	\	\	\	\	\	\	\	\
ER	Moldova	16					\	\	\	\	\	\	\	\	\
ES	Estonia	15					\	\	\	\	\	\	\	\	\
ET	Ethiopia	37					\	\	\	\	\	\	\	\	\
EV	Belarus	16					\	\	\	\	\	\	\	\	\
EX	Kyrgyzstan	17					\	\	\	\	\	\	\	\	\
EY	Tajikistan	17					\	\	\	\	\	\	\	\	\
EZ	Turkmenistan	17					\	\	\	\	\	\	\	\	\

A-19 This is not a prefix list, See Appendix A12 for oddball prefixes

Prefix	Entity	CQ	Mix	SSB	CW	Data	80	40	30	20	17	15	12	10	
F	France	14					\	\	\	\	\	\	\	\	\
FG	Guadeloupe	8					\	\	\	\	\	\	\	\	\
FH	Mayotte	39					\	\	\	\	\	\	\	\	\
FJ	St Barthelemy	8					\	\	\	\	\	\	\	\	\
FK	Chesterfield Is	30					\	\	\	\	\	\	\	\	\
FK	New Caledonia	32					\	\	\	\	\	\	\	\	\
FM	Martinique	8					\	\	\	\	\	\	\	\	\
FO	French Poly	32					\	\	\	\	\	\	\	\	\
FO0	Austral Is	32					\	\	\	\	\	\	\	\	\
FO0	Clipperton Is	7					\	\	\	\	\	\	\	\	\
FO0	Marquesas Is	31					\	\	\	\	\	\	\	\	\
FP	St Pierre & Miquelon	5					\	\	\	\	\	\	\	\	\
FR	Reunion	39					\	\	\	\	\	\	\	\	\
FR/G	Glorioso Is	39					\	\	\	\	\	\	\	\	\
FR/J	Juan de Nova, Europa	39					\	\	\	\	\	\	\	\	\
FR/T	Tromelin	39					\	\	\	\	\	\	\	\	\
FS	St Martin	8					\	\	\	\	\	\	\	\	\
FT_W	Crozet	39					\	\	\	\	\	\	\	\	\
FT_X	Kerguelen Is	39					\	\	\	\	\	\	\	\	\
FT_Z	Amsterdam	39					\	\	\	\	\	\	\	\	\
FW	Wallis	32					\	\	\	\	\	\	\	\	\
FY	Fr Guiana	9					\	\	\	\	\	\	\	\	\
G	England	14					\	\	\	\	\	\	\	\	\
GD	Isle of Man	14					\	\	\	\	\	\	\	\	\
GI	No Ireland	14					\	\	\	\	\	\	\	\	\
GJ	Jersey	14					\	\	\	\	\	\	\	\	\
GM	Scotland	14					\	\	\	\	\	\	\	\	\
GU	Guernsey	14					\	\	\	\	\	\	\	\	\
GW	Wales	14					\	\	\	\	\	\	\	\	\
H4	Solomon Is	28					\	\	\	\	\	\	\	\	\
H40	Temotu	28					\	\	\	\	\	\	\	\	\
HA	Hungary	15					\	\	\	\	\	\	\	\	\
HB	Switzerland	14					\	\	\	\	\	\	\	\	\
HB0	Liechtenstein	14					\	\	\	\	\	\	\	\	\
HC	Ecuador	10					\	\	\	\	\	\	\	\	\
HC8	Galapagos Is	10					\	\	\	\	\	\	\	\	\
HH	Haiti	8					\	\	\	\	\	\	\	\	\
HI	Dominican R	8					\	\	\	\	\	\	\	\	\
HK	Colombia	9					\	\	\	\	\	\	\	\	\
HK0	Malpelo Is	9					\	\	\	\	\	\	\	\	\
HK0	San Andres	7					\	\	\	\	\	\	\	\	\
HL	So Korea	25					\	\	\	\	\	\	\	\	\
HP	Panama	7					\	\	\	\	\	\	\	\	\
HR	Honduras	7					\	\	\	\	\	\	\	\	\
HS	Thailand	26					\	\	\	\	\	\	\	\	\
HV	Vatican	15					\	\	\	\	\	\	\	\	\
HZ	Saudi Arabia	21					\	\	\	\	\	\	\	\	\
I	Italy	33					\	\	\	\	\	\	\	\	\
IS	Sardinia	15					\	\	\	\	\	\	\	\	\
J2	Djibouti	37					\	\	\	\	\	\	\	\	\
J3	Grenada	8					\	\	\	\	\	\	\	\	\
J5	Guinea-Bissau	35					\	\	\	\	\	\	\	\	\
J6	St Lucia	8					\	\	\	\	\	\	\	\	\
J7	Dominica	8					\	\	\	\	\	\	\	\	\
J8	St Vincent	8					\	\	\	\	\	\	\	\	\
JA	Japan	25					\	\	\	\	\	\	\	\	\
JD	Minami Tori	27					\	\	\	\	\	\	\	\	\

A-19 This is not a prefix list, See Appendix A12 for oddball prefixes

Prefix	Entity	CQ	Mix	SSB	CW	Data	80	40	30	20	17	15	12	10	
JD	Ogasawara	27					\	\	\	\	\	\	\	\	\
JT	Mongolia	23					\	\	\	\	\	\	\	\	\
JW	Svalbard	40					\	\	\	\	\	\	\	\	\
JX	Jan Mayen	40					\	\	\	\	\	\	\	\	\
JY	Jordan	20					\	\	\	\	\	\	\	\	\
K	United States						\	\	\	\	\	\	\	\	\
KG4	Guantanamo Bay	8					\	\	\	\	\	\	\	\	\
KH0	Mariana Is	27					\	\	\	\	\	\	\	\	\
KH1	Baker & Howland Is	31					\	\	\	\	\	\	\	\	\
KH2	Guam	27					\	\	\	\	\	\	\	\	\
KH3	Johnston Is	31					\	\	\	\	\	\	\	\	\
KH4	Midway Is	31					\	\	\	\	\	\	\	\	\
KH5	Palmyra, Jarvis Is	31					\	\	\	\	\	\	\	\	\
KH5K	Kingman Reef	31					\	\	\	\	\	\	\	\	\
KH6	Hawaii	31					\	\	\	\	\	\	\	\	\
KH7K	Kure Is	31					\	\	\	\	\	\	\	\	\
KH8	Swains Island	32					\	\	\	\	\	\	\	\	\
KH8	Am Samoa	32					\	\	\	\	\	\	\	\	\
KH9	Wake Is	31					\	\	\	\	\	\	\	\	\
KL	Alaska	1					\	\	\	\	\	\	\	\	\
KP1	Navassa Is	8					\	\	\	\	\	\	\	\	\
KP2	Virgin Is	8					\	\	\	\	\	\	\	\	\
KP4	Puerto Rico	8					\	\	\	\	\	\	\	\	\
KP5	Desecheo Is	8					\	\	\	\	\	\	\	\	\
LA	Norway	14					\	\	\	\	\	\	\	\	\
LU	Argentina	13					\	\	\	\	\	\	\	\	\
LX	Luxembourg	14					\	\	\	\	\	\	\	\	\
LY	Lithuania	15					\	\	\	\	\	\	\	\	\
LZ	Bulgaria	20					\	\	\	\	\	\	\	\	\
OA	Peru	10					\	\	\	\	\	\	\	\	\
OD	Lebanon	20					\	\	\	\	\	\	\	\	\
OE	Austria	15					\	\	\	\	\	\	\	\	\
OH	Finland	15					\	\	\	\	\	\	\	\	\
OH0	Aland Is	15					\	\	\	\	\	\	\	\	\
OJ0	Market Reef	15					\	\	\	\	\	\	\	\	\
OK	Czech Republic	15					\	\	\	\	\	\	\	\	\
OM	Slovakia	15					\	\	\	\	\	\	\	\	\
ON	Belgium	14					\	\	\	\	\	\	\	\	\
OX	Greenland	40					\	\	\	\	\	\	\	\	\
OY	Faroe Is	14					\	\	\	\	\	\	\	\	\
OZ	Denmark	14					\	\	\	\	\	\	\	\	\
P2	Papua New Guinea	28					\	\	\	\	\	\	\	\	\
P4	Aruba	9					\	\	\	\	\	\	\	\	\
P5	No Korea	25					\	\	\	\	\	\	\	\	\
PA	Netherlands	14					\	\	\	\	\	\	\	\	\
PJ2	Neth Antilles	9					\	\	\	\	\	\	\	\	\
PJ5	St Maarten	8					\	\	\	\	\	\	\	\	\
PY	Brazil	11					\	\	\	\	\	\	\	\	\
PY0F	Fernando de Noronha	11					\	\	\	\	\	\	\	\	\
PY0P	St Peter/Paul Rocks	11					\	\	\	\	\	\	\	\	\
PY0T	Trindade	11					\	\	\	\	\	\	\	\	\
PZ	Surinam	9					\	\	\	\	\	\	\	\	\
R1FJ	Franz Josef Land	40					\	\	\	\	\	\	\	\	\
R1MV	Malyj Vysotskij Is	16					\	\	\	\	\	\	\	\	\
S0	Western Sahara	33					\	\	\	\	\	\	\	\	\
S2	Bangladesh	22					\	\	\	\	\	\	\	\	\
S5	Slovenia	15					\	\	\	\	\	\	\	\	\

Prefix	Entity	CQ	Mix	SSB	CW	Data	80	40	30	20	17	15	12	10	
S7	Seychelles	39					\	\	\	\	\	\	\	\	\
S9	Sao Tome	36					\	\	\	\	\	\	\	\	\
SM	Sweden	14					\	\	\	\	\	\	\	\	\
SP	Poland	15					\	\	\	\	\	\	\	\	\
ST	Sudan	34					\	\	\	\	\	\	\	\	\
SU	Egypt	34					\	\	\	\	\	\	\	\	\
SV	Greece	20					\	\	\	\	\	\	\	\	\
SV5	Dodecanese	20					\	\	\	\	\	\	\	\	\
SV9	Crete	20					\	\	\	\	\	\	\	\	\
SV9	Mt Athos	20					\	\	\	\	\	\	\	\	\
T2	Tuvalu	31					\	\	\	\	\	\	\	\	\
T30	West Kiribati	31					\	\	\	\	\	\	\	\	\
T31	Central Kiribati	31					\	\	\	\	\	\	\	\	\
T32	East Kiribati	31					\	\	\	\	\	\	\	\	\
T33	Banaba	31					\	\	\	\	\	\	\	\	\
T5	Somalia	37					\	\	\	\	\	\	\	\	\
T7	San Marino	15					\	\	\	\	\	\	\	\	\
T8	Belau	27					\	\	\	\	\	\	\	\	\
T9	Bosnia Now E7	15					\	\	\	\	\	\	\	\	\
TA	Turkey	20					\	\	\	\	\	\	\	\	\
TF	Iceland	40					\	\	\	\	\	\	\	\	\
TG	Guatemala	7					\	\	\	\	\	\	\	\	\
TI	Costa Rica	7					\	\	\	\	\	\	\	\	\
TI9	Cocos Is	7					\	\	\	\	\	\	\	\	\
TJ	Cameroon	36					\	\	\	\	\	\	\	\	\
TK	Corsica	15					\	\	\	\	\	\	\	\	\
TL	Cen African Rep	36					\	\	\	\	\	\	\	\	\
TN	Congo	36					\	\	\	\	\	\	\	\	\
TR	Gabon	36					\	\	\	\	\	\	\	\	\
TT	Chad	36					\	\	\	\	\	\	\	\	\
TU	Ivory Coast	35					\	\	\	\	\	\	\	\	\
TY	Benin	35					\	\	\	\	\	\	\	\	\
TZ	Mali	35					\	\	\	\	\	\	\	\	\
UA	Russia	16					\	\	\	\	\	\	\	\	\
UA2	Kaliningrad	15					\	\	\	\	\	\	\	\	\
UA9	Russia (Asiatic)	23					\	\	\	\	\	\	\	\	\
UK	Uzbekistan	17					\	\	\	\	\	\	\	\	\
UN	Kazakhstan	17					\	\	\	\	\	\	\	\	\
UR	Ukraine	16					\	\	\	\	\	\	\	\	\
V2	Antigua, Barbuda	8					\	\	\	\	\	\	\	\	\
V3	Belize	7					\	\	\	\	\	\	\	\	\
V4	St Kitts, Nevis	8					\	\	\	\	\	\	\	\	\
V5	Namibia	38					\	\	\	\	\	\	\	\	\
V6	Micronesia	27					\	\	\	\	\	\	\	\	\
V7	Marshall Is	31					\	\	\	\	\	\	\	\	\
V8	Brunei	28					\	\	\	\	\	\	\	\	\
VE	Canada						\	\	\	\	\	\	\	\	\
VK	Australia	30					\	\	\	\	\	\	\	\	\
VK0	Heard Is	39					\	\	\	\	\	\	\	\	\
VK0	Macquarie Is	30					\	\	\	\	\	\	\	\	\
VK9C	Cocos-Keeling	29					\	\	\	\	\	\	\	\	\
VK9L	Lord Howe Is	30					\	\	\	\	\	\	\	\	\
VK9M	Mellish Reef	30					\	\	\	\	\	\	\	\	\
VK9N	Norfolk Is	32					\	\	\	\	\	\	\	\	\
VK9W	Willis Is	30					\	\	\	\	\	\	\	\	\
VK9X	Christmas Is	29					\	\	\	\	\	\	\	\	\

Prefix	Entity	CQ	Mix	SSB	CW	Data	80	40	30	20	17	15	12	10	
VP2E	Anguilla	8					\	\	\	\	\	\	\	\	\
VP2M	Montserrat	8					\	\	\	\	\	\	\	\	\
VP2V	Br Virgin Is	8					\	\	\	\	\	\	\	\	\
VP5	Turks & Caicos Is	8					\	\	\	\	\	\	\	\	\
VP6D	Ducie Island	32					\	\	\	\	\	\	\	\	\
VP6	Pitcairn Is	32					\	\	\	\	\	\	\	\	\
VP8	Falkland Is	13					\	\	\	\	\	\	\	\	\
VP8	So Georgia Is	13					\	\	\	\	\	\	\	\	\
VP8	So Orkney Is	13					\	\	\	\	\	\	\	\	\
VP8	So Sandwich Is	13					\	\	\	\	\	\	\	\	\
VP8	So Shetland Is	13					\	\	\	\	\	\	\	\	\
VP9	Bermuda	5					\	\	\	\	\	\	\	\	\
VQ9	Chagos	39					\	\	\	\	\	\	\	\	\
VR2	Hong Kong	24					\	\	\	\	\	\	\	\	\
VU	India	22					\	\	\	\	\	\	\	\	\
VU7	Andaman & Nicobar Is	26					\	\	\	\	\	\	\	\	\
VU7	Lakshadweep Is	22					\	\	\	\	\	\	\	\	\
XE	Mexico	6					\	\	\	\	\	\	\	\	\
XF4	Revilla Gigedo	6					\	\	\	\	\	\	\	\	\
XT	Burkina Faso	35					\	\	\	\	\	\	\	\	\
XU	Cambodia	26					\	\	\	\	\	\	\	\	\
XW	Laos	26					\	\	\	\	\	\	\	\	\
XX9	Macao	24					\	\	\	\	\	\	\	\	\
XZ	Myanmar (Burma)	26					\	\	\	\	\	\	\	\	\
YA	Afghanistan	21					\	\	\	\	\	\	\	\	\
YB	Indonesia	28					\	\	\	\	\	\	\	\	\
YI	Iraq	21					\	\	\	\	\	\	\	\	\
YJ	Vanuatu	32					\	\	\	\	\	\	\	\	\
YK	Syria	20					\	\	\	\	\	\	\	\	\
YL	Latvia	15					\	\	\	\	\	\	\	\	\
YN	Nicaragua	7					\	\	\	\	\	\	\	\	\
YO	Romania	20					\	\	\	\	\	\	\	\	\
YS	El Salvador	7					\	\	\	\	\	\	\	\	\
YTYU	Serbia	15					\	\	\	\	\	\	\	\	\
YV	Venezuela	9					\	\	\	\	\	\	\	\	\
YV0	Aves Is	8					\	\	\	\	\	\	\	\	\
Z2	Zimbabwe	38					\	\	\	\	\	\	\	\	\
Z3	Macedonia	15					\	\	\	\	\	\	\	\	\
ZA	Albania	15					\	\	\	\	\	\	\	\	\
ZB	Gibraltar	14					\	\	\	\	\	\	\	\	\
ZC	Cyprus SBA	20					\	\	\	\	\	\	\	\	\
ZD7	St Helena	36					\	\	\	\	\	\	\	\	\
ZD8	Ascension Is	36					\	\	\	\	\	\	\	\	\
ZD9	Tristan da Cunha	38					\	\	\	\	\	\	\	\	\
ZF	Cayman Is	8					\	\	\	\	\	\	\	\	\
ZK1	No Cook Is	32					\	\	\	\	\	\	\	\	\
ZK1	So Cook Is	32					\	\	\	\	\	\	\	\	\
ZK2	Niue	32					\	\	\	\	\	\	\	\	\
ZK3	Tokelau Is	31					\	\	\	\	\	\	\	\	\
ZL	New Zealand	32					\	\	\	\	\	\	\	\	\
ZL7	Chatham Is	32					\	\	\	\	\	\	\	\	\
ZL8	Kermadec Is	32					\	\	\	\	\	\	\	\	\
ZL9	Auckland, Campbell Is	32					\	\	\	\	\	\	\	\	\
ZP	Paraguay	11					\	\	\	\	\	\	\	\	\
ZS	So Africa	38					\	\	\	\	\	\	\	\	\
ZS8	Pr Edward & Marion Is	38					\	\	\	\	\	\	\	\	\

A-19 This is not a prefix list, See Appendix A12 for oddball prefixes

**APPENDIX A12 DXCC
COUNTRIES - Difficulty
Most Wanted Countries**

The following are among the
most difficult 100 countries to
work.

1A0 SMOM
1S Spratly
3B7 Agalega & StBrandon
3B9 Rodriguez
3C Equatorial Guinea
3C0 Annobon
3D2C Conway Reef
3D2R Rotuma
3V Tunisia
3W Vietnam
3X Guinea
3Y Bouvet
3Y Peter I
5A Libya
5R8 Madagascar
5X Uganda
7O Yemen
8Q Maldive
9N Nepal
9U Burundi
A5 Bhutan
A6 United Arab Emirates
BS7H Scarborough Reef
BV9P Pratas
CE0X San Felix
CE0Z Juan Fernandez
CY0 Sable Island
CY9 St Paul Island
D6 Comoros
E3 Eritrea
EP Iran
ET Ethiopia
FO/C Clipperton
FR/G Glorioso
FR/J Juan De Nova & Europa
FR/T Tromelin
FT5W Crozet
FT5X Kerguelen
FT5Z Amsterdam & St Paul
FW Wallis & Futuna
HK0/M Malpelo
JD1 Minami Torishima
JD1 Ogasawara
JX Jan Mayen
KH1 Baker & Howland
KH4 Midway

KH5 Palmyra & Jarvis
KH5K Kingman Reef
KH7K Kure
KH8 Swains Is (New DXCC
Entity)
KH9 Wake
KP1 Navassa
KP5 Desecheo Island
P5 North Korea
PY0S St Peter And Paul Rocks
PY0T Trindade & MartinVaz
R1M Malyj-Vysotskij
S0 Western Sahara
S2 Bangladesh
ST Sudan
SV/A Mount Athos
T2 Tuvalu
T31 Central Kiribati
T33 Banaba
T5 Somalia
T8 Palau
TI9 Cocos
TN Congo
TT8 Chad
TY Benin
VK0/H Heard
VK0/M Macquarie
VK9/W Willis
VK9/X Christmas
VK9/Y Cocos-Keeling
VK9L Lord Howe Island
VK9M Mellish Reef
VP8 South Orkneys
VP8 South Shetlands
VP8/G South Georgia
VP8/S South Sandwich
VU4 Andaman & Nicobar
VU7 Lakshadweep
XU Cambodia
XW Laos
XZ Myanmar
YA Afghanistan
YI Iraq
YK Syria
YU Montenegro (New Entity)
YV0 Aves
ZC4 UK Bases on Cyprus
ZD9 Tristan Da Cunha & Gough
ZK1 North Cook
ZK3 Tokelau
ZL8 Kermadec
ZL9 Auckland & Campbell
ZS8 Prince Edward & Marion

**Countries with ham
populations over 1,000**
Argentina
Australia
Austria
Balearic Islands
Belgium
Bolivia
Brazil
Canada
Canary Islands
Chile
Colombia
Czech. Republic
Denmark
Ecuador
El Salvador
England
Fed. Rep. Of Germany
Finland
France
Greece
Hungary
India
Indonesia
Ireland
Israel
Italy
Japan
Lithuania
Mexico
Netherlands
New Zealand
Northern Ireland
Norway
Panama
Paraguay
Peru
Philippines
Poland
Portugal
Republic of South Africa
Romania
Russian Federation
Scotland
Slovak Republic
Slovenia
South Korea
Spain
Sweden
Taiwan
Ukraine
Uruguay
USA

COUNTRIES OFTEN EASY TO WORK IN CONTESTS

Aland Islands
Anguilla
Antigua/Barbuda
Aruba
Bahamas
Barbados
Belize
Bermuda
Bulgaria
Cayman is.
China
Croatia
Estonia
Georgia
Grenada
Guernsey
Hong Kong
Iceland
Isle of Man
Latvia
Lebanon
Luxembourg
Macedonia
Moldova
Morocco
Netherlands Antilles
Nicaragua
Saint Martin
Sardinia
Singapore
St Lucia
St Martin
St Vincent
Switzerland

COUNTRIES NOT SERVED BY THE ARRL OUGOING QSL BUREAU

Afghanistan
American Samoa
Angola
Baker and Howland Islands
Benin
Bhutan
Burundi
Cameroon
Central African Republic
Chad
Congo
Desecheo Island
Egypt
Equatorial Guinea
Guinea
Guinea-Bissau
Kampuchea
Kiribati
Kure Islands
Laos
Lesotho
Libya
Madagascar
Malawi
Maldive

AC6V DX Reference Guide
Index

NOTE: A complete guide to Ham Jargon and terms can be found at URL:
http://ac6v.com/jargon.htm